Extension Innovation Method

Systems Evaluation, Prediction, and Decision-Making Series

Series Editor
Yi Lin, PhD
Professor of Systems Science and Economics
School of Economics and Management
Nanjing University of Aeronautics and Astronautics

Capital Account Liberation: Methods and Applications
Ying Yirong and Jeffrey Yi-Lin Forrest
ISBN 978-1-4987-1226-2

Grey Game Theory and Its Applications in Economic Decision-Making
Zhigeng Fang, Sifeng Liu, Hongxing Shi, and Yi Lin
ISBN 978-1-4200-8739-0

Hybrid Rough Sets and Applications in Uncertain Decision-Making
Lirong Jian, Sifeng Liu, and Yi Lin
ISBN 978-1-4200-8748-2

Introduction to Theory of Control in Organizations
Vladimir N. Burkov, Mikhail Goubko, Nikolay Korgin, and Dmitry Novikov
ISBN 978-1-4987-1423-5

Investment and Employment Opportunities in China
Yi Lin and Tao Lixin
ISBN 978-1-4822-5207-1

Irregularities and Prediction of Major Disasters
Yi Lin
ISBN: 978-1-4200-8745-1

Measurement Data Modeling and Parameter Estimation
Zhengming Wang, Dongyun Yi, Xiaojun Duan, Jing Yao, and Defeng Gu
ISBN 978-1-4398-5378-8

Optimization of Regional Industrial Structures and Applications
Yaoguo Dang, Sifeng Liu, and Yuhong Wang
ISBN 978-1-4200-8747-5

Systems Evaluation: Methods, Models, and Applications
Sifeng Liu, Naiming Xie, Chaoqing Yuan, and Zhigeng Fang
ISBN 978-1-4200-8846-5

Systemic Yoyos: Some Impacts of the Second Dimension
Yi Lin
ISBN 978-1-4200-8820-5

Theory and Approaches of Unascertained Group Decision-Making
Jianjun Zhu
ISBN 978-1-4200-8750-5

Theory of Science and Technology Transfer and Applications
Sifeng Liu, Zhigeng Fang, Hongxing Shi, and Benhai Guo
ISBN 978-1-4200-8741-3

Extension Innovation Method

Chunyan Yang

CRC Press
Taylor & Francis Group
Boca Raton London New York

CRC Press is an imprint of the
Taylor & Francis Group, an **informa** business

CRC Press
Taylor & Francis Group
6000 Broken Sound Parkway NW, Suite 300
Boca Raton, FL 33487-2742

International Standard Book Number-13: 978-1-138-36721-0 (Hardback)

Visit the Taylor & Francis Web site at
http://www.taylorandfrancis.com

and the CRC Press Web site at
http://www.crcpress.com

MIX
Paper from
responsible sources
FSC   FSC® C013056
www.fsc.org

Printed and bound in Great Britain by
TJ International Ltd, Padstow, Cornwall

About the Book

The core of innovation consists of creative ideas. Methods of innovation provide effective means for independent innovation. As described by Extenics, a discipline that originated in China, the generation of creative ideas inevitably abides by certain rules and approaches!

Extension innovation method is an approach used for generating creative ideas. It utilizes basic theories of Extenics to establish a combined modeling-quantification method that can be learned effortlessly and operated conveniently. In addition, this method is able to indicate where one can start her innovation, identify the basis of creative idea generation, suggest tools for creative idea generation, and illuminate how to evaluate creative ideas and select the superior creative ideas among available choices, etc. This method is also applied to elicit new ideas and resolving contradictory problems in various fields. In this book, all of the commonly used extension innovation methods are introduced and analyzed thoroughly to demonstrate that they are practically applicable. This book makes it convenient for readers of different levels and different knowledge backgrounds to study this set of innovation tools. Moreover, many highly accessible cases are provided in each section of this book.

This book is suitable for engineers, managerial decision makers, and university scholars and students who are learning about product innovation, technological innovation, management innovation, system innovation, etc. This book can be used as a textbook for elective courses for undergraduates, graduate students, and doctoral students studying relevant specialties that involve innovation or innovative activities.

Contents

Preface

Time flies. Twenty-two years have passed by in a flash since I was first professionally occupied in theoretical and applied research on Extenics, a discipline originated in China. In this those years, Extenics evolved in various up- and-down moments, and I witnessed its growth and development.

With the continuous development of theoretical studies on Extenics, it becomes increasingly urgent to investigate its applications so that it can be popularized and promoted. In addition, the tasks of internationalization and socialization of the theory become heavier day by day. Against this backdrop, the Extenics research team keeps expanding, and the number of researchers who make use of extension innovation methods in diverse disciplines rises constantly. Therefore, there is an urgent need to publish a book that is dedicated to the introduction of extension innovation methods and provides a convenient way for people to study and apply the methods.

The purpose of writing this book is to share with beginners the basic operational methods that can be used for innovation purposes or for solving contradictory problems. The relevant theoretical knowledge will be introduced briefly as prerequisites, and readers who are interested in learning more about the theory of Extenics can reference the monograph *Extenics: Theory, Method and Application* published by Science Press of China and The Educational Publisher Inc. of U.S.A in 2013 and the monograph *Extenics* (in Chinese) published by Science Press of China in 2014.

Chapter 1 of this book briefly introduces profiles of the extension innovation method. In Chapter 2, a starting point of innovation is presented, i.e. the extension modeling method. The extensible analysis method, one of two creative idea generation bases, is described in Chapter 3, and conjugate analysis, the other method, is illustrated in Chapter 4. In Chapter 5, a tool that can be utilized to generate creative ideas is presented: extension transformation. For the evaluation and optimum selection of creative ideas, Chapter 6 introduces the superiority evaluation method. In Chapters 7 and 8, extension creative idea generation methods for solving contradictory problems and creating a new product are provided, respectively. In each section, several case studies are presented to assist readers in comprehending various methods and their applications. Each chapter also includes reflections and exercises.

This work is supported by NSFC fund projects (61273306, 61503085) and Science and Technology Planning Projects of Guangdong Province (2012B061000012, 2016A040404015). My hope is that this book can provide tools of innovation and solving contradictory problems for teaching and scientific research personnel in universities and research & development institutions of businesses, as well as product innovation staff and managers in various enterprises. I expect that they will be able to apply these approaches in their own research fields and thus put forward more extension innovation methods appropriate for all professions.

I appreciate the substantial support given in the process of writing this book by Prof. Wen Cai, who is the founder of Extenics and a reviewer of this book. Undeniably, he has provided many valuable comments and suggestions. I also express my gratitude to the National Natural Science Foundation of China and Guangdong Provincial Department of Science and Technology, both of which offer me strong support for my research. I am also grateful to Guangdong University of Technology (GDUT) for the free and democratic scientific research environment it provides. My thanks also go to Dr. Long Tang and Prof. Xingsen Li from the Research Institute of Extenics and Innovation Methods, GDUT; Prof. Weihua Li from the School of Computers, GDUT; Prof. Jeffrey Y-L Forrest from Slippery Rock University; Dr. Haolan Zhang from Ningbo Institute of Technology, Zhejiang University—they have all energetically supported me in terms of the translation and proofreading on this book. My gratitude also goes to my graduate students, Zhiming Li, Ningning Qi, Yongqiang Liao, and Liangwei Luo, for several case studies that they provided. Last but not least, I also give my grateful thanks to Dr. Xiaomei Li and Dr. Jintao Yang from the School of Computers, GDUT, who have cooperated closely with me for so many years.

In the publication of this book, I would like to express my special thanks to my husband, Prof. Yongjie Wang, who has been with me for thirty years and has given me unlimited tolerance and support. It is with his strong support that I have the courage to face the setbacks in my career, which has led to fruitful results.

Due to the limitations of my knowledge, omissions and even mistakes are unavoidable; I implore readers to criticize.

Chunyan Yang

Foreword

Human history is the history of solving contradictions and making constant innovations. It is of great significance for human intelligence to analyze the possibility of expanding matters and affairs and rules of innovation with formalized models and to summarize methods of solving problems. It is of great value for improving machine intelligence to discuss theories and methods of solving contradictory problems according to these research achievements. Research on Extenics is conducted for the sake of these objectives.

The study of Extenics started in 1976, and the first paper on the subject, "Extension Set and Incompatible Problems," was published in 1983. In the past decade, with the efforts made by Extenics researchers, the framework of Extenics has been gradually formed, the method system of extension innovation has been established, research in several fields has been carried out, and the profile of a new discipline has been formed.

Quite a few scholars have joined in the construction of this new discipline in recent years. A number of books on Extenics are in urgent demand as references for researchers who seek to apply, study, popularize, and promote Extenics.

The book *Extension Innovation Method*, written by Prof. Chunyan Yang, systematically introduces commonly used extension innovation methods and thus provides preliminary knowledge of Extenics, extension thinking methods, and research topics for many scholars who are studying Extenics.

Wen Cai

Expert with Special Country-level Contributions
Founder of Extenics, a New Discipline

Main Symbol Description

Symbol	Meaning
$M = (O_m, c_m, v_m)$	1-dimensional matter-element
$M(t) = (O_m(t), c_m, v_m(t))$	1-dimensional parametric matter-element
$A = (O_a, c_a, v_a)$	1-dimensional affair-element
$A(t) = (O_a(t), c_a, v_a(t))$	1-dimensional parametric affair-element
$M = \begin{bmatrix} O_m, & c_{m1}, & v_{m1} \\ & c_{m2}, & v_{m2} \\ & \vdots & \vdots \\ & c_{mn}, & v_{mn} \end{bmatrix} = (O_m, C_m, V_m)$	n-dimensional matter-element
$A = \begin{bmatrix} O_a, & c_{a1}, & v_{a1} \\ & c_{a2}, & v_{a2} \\ & \vdots & \vdots \\ & c_{an}, & v_{an} \end{bmatrix} = (O_a, C_a, V_a)$	n-dimensional affair-element
$R = \begin{bmatrix} O_r, & c_{r1}, & v_{r1} \\ & c_{r2}, & v_{r2} \\ & \vdots, & \vdots \\ & c_{rn}, & v_{rn} \end{bmatrix} = (O_r, C_r, V_r)$	n-dimensional relation-element
$B = (O, c, v)$	1-dimensional basic-element

(c, v)	characteristic-element
$B(t) = (O(t), c, v(t))$	1-dimensional parametric basic-element
$B = (O, C, V) = \begin{bmatrix} Object, & c_1, & v_1 \\ & c_2, & v_2 \\ & \vdots & \vdots \\ & c_n, & v_n \end{bmatrix}$	n-dimensional basic-element
$B(t) = (O(t), C, V(t)) = \begin{bmatrix} O(t), & c_1, & v_1(t) \\ & c_2, & v_2(t) \\ & \vdots & \vdots \\ & c_n, & v_n(t) \end{bmatrix}$	n-dimensional parametric basic-element
$\{B\} = (\{O\}, \ C, \ V) = \begin{bmatrix} \{O\}, & c_1, & V_1 \\ & c_2, & V_2 \\ & \vdots & \vdots \\ & c_n, & V_n \end{bmatrix}$	n-dimensional class basic-element
$T\Gamma = \Gamma'$	substituting transformation
$T_1\Gamma = \Gamma \oplus \Gamma_1$	adding transformation
$T_2\Gamma = \Gamma \ominus \Gamma_1$	removing transformation
$T\Gamma = \alpha\Gamma, \alpha > 1$	enlarging transformation
$T\Gamma = \alpha\Gamma, 0 < \alpha < 1$	shrinking transformation
$T\Gamma = \{\Gamma_1, \Gamma_2, ..., \Gamma_n \,|\, \Gamma_1 \oplus \Gamma_2 \oplus \cdots \oplus \Gamma_n = \Gamma\}$	decomposing transformation
$T\Gamma = \{\Gamma, \Gamma^*\}$	duplicating transformation

T_φ	the first-order conductive transformation of the active transformation φ
$T_{\varphi^{(n)}}$	the n-order conductive transformation of the active transformation φ
$_{\Gamma_1}T_{\Gamma_2}$	the transformation of Γ_1 causes the conductive transformation of Γ_2
$\varphi \Rightarrow {}_0T_1 \Rightarrow {}_1T_2 \Rightarrow \cdots \Rightarrow {}_{n-2}T_{n-1} \Rightarrow {}_{n-1}T_n$	n-order conductive transformation of φ
T_2T_1	PRODUCT transformation of T_1 and T_2
T^{-1}	INVERSE transformation of T
$T_1 \wedge T_2$	AND transformation of T_1 and T_2
$T_1 \vee T_2$	OR transformation of T_1 and T_2
$c(\varphi) = c(B_0') - c(B_0)$	the active variable of φ about characteristic c for basic-element B_0
$c(T_\varphi) = c(B') - c(B)$	the conductive transformation of φ about characteristic c for basic-element B
$\tilde{E}(T)$	extension set
E_+	positive field
E_-	negative field
E_0	zero boundary
$E_+(T)$	positive extensible field (positive qualitative change field) of $\tilde{E}(T)$
$E_-(T)$	negative extensible field (negative qualitative change field) of $\tilde{E}(T)$

$E_+(T)$	positive stable field (positive quantitative change field) of $\tilde{E}(T)$
$E_-(T)$	negative stable field (negative quantitative change field) of $\tilde{E}(T)$
$E_0(T)$	extensible boundary of $\tilde{E}(T)$
$\tilde{E}(B)(T)$	basic-element extension set
$y = k(u)$	dependent function
$y' = T_k k(T_u u)$	extension function
$<a, b>$	the interval formed by a and b, it can indicate an open interval, a closed interval, or a half-open and half-closed interval
$\rho(x, x_0, X)$	extension distance between x and interval X about x_0
$D(x, x_0, X_0, X)$	the place value of point x_0 about the nest intervals composed by intervals X and X_0
$P = G * L, \ G \uparrow L$	incompatible problem
$P = (G_1 \wedge G_2) * L, \ (G_1 \wedge G_2) \uparrow L$	antithetical problem
$\text{re}(O_m)$	material part of matter O_m
$\text{im}(O_m)$	nonmaterial part of matter O_m
$\text{hr}(O_m)$	hard part of matter O_m
$\text{sf}(O_m)$	soft part of matter O_m
$\text{lt}(O_m)$	latent part of matter O_m
$\text{ap}(O_m)$	apparent part of matter O_m
$\text{ps}_c(O_m)$	positive part of matter O_m
$\text{ng}_c(O_m)$	negative part of matter O_m
$=$	equal

$\neq$	not equal
$\sim$	correlation
$\overset{\sim}{\rightarrow}$	directional correlation
$\Rightarrow$	implication
$\dashv$	divergence
$\oplus$	add
$\otimes$	product
$\ominus$	remove
$//$	decomposition
@	existence, realization
$\overline{@}$	not existence, not realization
$\wedge$	AND operation
$\vee$	OR operation
$\neg$	NOT operation
$A \dashv B$	A extends B
$\overline{B}$ or $\neg B$	NOT basic-element of B
$\overline{M}$ or $\neg M$	NOT matter-element of B
$\overline{A}$ or $\neg A$	NOT affair-element of B
$\overline{R}$ or $\neg R$	NOT relation-element of B

Chapter 1

Overview of Extension Innovation Method

Content Summary

Extenics is an original trans-disciplinary discipline originated by Professor Cai Wen, a Chinese scholar, in 1983. The theory examines the possibility of the extension and transformation of things, the rules and methods of innovation in a formalized approach. Moreover, the discipline is also used for solving contradictory problems, defined as problems that cannot be resolved within existing conditions.

The core of innovation lies in generating creative ideas, while the innovation method represents the fundamental source of independent innovation. Extenics tells us: Creative ideas can be generated based on their rules by using specific methods!

As a method for generating creative ideas, based on the basic theory of Extenics, extension innovation method can help establish a convenient, easily learned, and user-friendly method by combining modeling with quantification, and it can indicate to one where the starting point of innovation is, where the base for generating creative ideas is located, which tools should be adopted for generating creative ideas, and how we can evaluate creative ideas, etc., all of which can be beautifully applied in innovation and in solving contradictory problems in all theoretical and applied fields.

1.1 Introduction to Extenics

Extenics consists of the extension theory, the extension innovation method and the extension engineering. The extension theory at its core aims to solve contradictory problems in the real world. The extension innovation methodology (also known as the extension methodology) incorporates the method system and the extension logic as the logic foundation; Extenics forms extension engineering by integrating all fields in holistically.

The discipline framework of Extenics is shown in Figure 1.1.1.

Extenics is a newly rising discipline across different disciplines such as mathematics, philosophy, and engineering. It is trans-disciplinary, covering a wide range fields, including cybernetics, information theory, and systems theory. Where there are quantitative relations and spatial forms, there is the existence of mathematics; where there exist contradictory problems, there is a favorable position for the use of Extenics. The application effectiveness of Extenics in various disciplines and engineering fields lies in providing new ideas and methods instead of discovering a new experimental fact.

In 1998, a paper entitled "Review on Extension Engineering Method" published in the *Chinese Science Bulletin*, an authoritative journal of the Chinese Academy of Sciences, pointed out that: "Extenics is a new discipline full of vitalities; its creation is the pride of the Chinese people; it belongs to not only China, but also the world." In 1999, "The Extension Theory and Its Application" was published. In 2013, "Review on Extenics" and a special review of "Basic Theory and Method System of Extenics" were published to comprehensively introduce Extenics. "Extension Theory and Its Application" has been formally certified by the certifying commission with

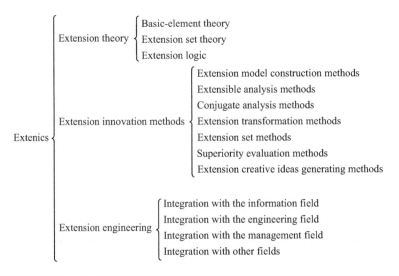

Figure 1.1.1   The discipline framework of Extenics.

Wu Wenjun, academician from the Chinese Academy of Sciences, as the director and Li Youping, academician from the Chinese Academy of Engineering, as the deputy director. The certification states that "Professor Cai Wen et al. have established a new discipline—Extenics—across the disciplines of philosophy, mathematics, and engineering after years of continuous research. Extenics is an original discipline, with far reaching values, that is established by scientists of our country."

In 2011, the work titled "Extension Theory and Its Application" was given the first prize of innovation award in the inaugural Wu Wenjun Artificial Intelligence Science and Technology Award. The Achievement Management Office of the People's Republic of China Ministry of Science and Technology officially announced the achievement to the world, stating: "The project, as an original innovative research, takes a leading and guiding position in similar research at home and abroad." At present, scholars in various fields establish results through investigating problems in their respective professional fields by employing the methods contained in this project. Many scholars from around the world have come to China to learn about this project and then bring the news and knowledge to different parts of the world. Extenics has been applied in research in various fields, showing broad application prospects.

"The founding and development of Extenics, the original discipline, indicate that the Chinese have the ability to carry out original innovation research; the National Natural Science Foundation is a strong guarantee to support the original innovative research", according to an information release from the National Natural Science Foundation.

People's Daily, Guangming Daily, and *Science Times*, among others, have introduced Extenics to the public successively with great details (please see the website of Extenics: http://Extenics.gdut.edu.cn/).

Regarding theoretical research, Extenics researchers have gradually established a theoretical system and methodology system for Extenics since 1983, publishing seventeen monographs (including "Extenics Series"), such as "Extenics, Extension Engineering" and "Extension Innovation Method" (in Chinese) as well as a number of papers and monographs in both English and traditional Chinese. An English monograph, entitled "Extenics: Theory, Method and Application" and published jointly by Science Press (P.R. China) and The Educational Publishing, Inc. (USA), has been preserved by the Library of Congress and is regarded as an instructional book for the five sessions of international research scholars on Extenics.

Regarding application research, the Chinese Association for Science and Technology (CAST) published achievements of Extenics application research on computer, management, control and detection, etc., in the "Discipline Development Report" in 2008 and 2010, respectively. Several monographs and papers, such as *Extension Strategy Generating System, Extension Set and Extension Data Mining, Extension Marketing Theory, Extension Strategy-Tactics-Planning (ESTP), Extension Data Mining Method and Its Computer Implementation*, and *Extension Design*, were included. Moreover, Extenics researchers have also successfully applied for patents and developed a number of extension softwares.

According to our incomplete statistics, as of 2016, seventy-six theoretical research and application research projects on or based on Extenics have been funded by the National Natural Science Foundation. There have been more than forty projects in various fields and professions that apply Extenics in their problem solving. In total, over 4,700 papers on Extenics or applications of Extenics have been published by Chinese journals, with about 1,500 doctoral dissertations and master's theses on Extenics in China.

Regarding team construction, an Extenics research team covering more than twenty provinces and cities as well as foreign countries has been developed. The Extenics Society of Chinese Association for Artificial Intelligence was established with the approval of the Ministry of Civil Affairs. Also, the Research Institute of Extenics and Innovation Methods of Guangdong University of Technology has been built to engage in Extenics projects. Since 1993, the Research Institute has recruited Extenics research scholars in China and from abroad to study Extenics and to cultivate Extenics research backbones. To date, nineteen sessions of domestic research scholars and five sessions of international scholars have been recruited. International scholars have included professors, PhD candidates, and engineers from the United States, India, Romania, and other countries. American professors have published two monographs on Extenics in the United States after finishing their study in China and returning home. Achievements completed jointly by professors from the Romanian Academy of Sciences and our institute were awarded the top prize at the Geneva International Invention Fair and have received the Russian Federation Award.

With regard to international academic exchanges, Extenics researchers have visited Spain, Romania, France, the United States, Italy, Germany, Britain, Australia, Brazil, India, South Korea, and Japan to introduce Extenics; they have visited Taiwan three times to organize Extenics workshops and extension innovation method seminars, they have visited Hong Kong several times to introduce Extenics and organize training class for the extension innovation method at several universities and industrial circles. The First International Symposium on Extenics and Innovation Methods, held in Beijing in August 2013, was a great success. Extenics is moving towards overseas.

Extenics, originating from one person's academic thought and a paper published more than thirty years ago, has been developed into a theory with a mature framework. In this period, a considerable number of scholars have participated in the construction of this new discipline. Extenics is expected to play an active role in world economic and social development along with the deepening of scientific and applied research.

The Extenics research has gone through two stages: the proposal of concepts and ideas, and the establishment of basic theoretical frameworks upon years of efforts. Extenics is now entering the next stage, in which application research and theory research are combined. To achieve this, however, a large quantity of arduous and rigorous work needs to be done to make Extenics become a mature, powerful discipline ready to produce beneficial applications.

Extenics is an original Chinese discipline and is gradually expanding beyond China to the world. As of this writing, China is still in a leading position, representing the current research level worldwide on Extenics and the latest international progress. If we continue to strengthen the research on the Extenics, we'd like to achieve breakthrough technical achievements ahead of the rest of the world.

1.2 The System and Essential Characteristics of Extension Innovation Method

Extenics proposes a new methodology to help people analyze the real world and solve contradictory problems in the real world with new perspectives, forming an extension innovation method system, as shown in Figure 1.2.1.

The basic characteristics of the method system consist of formalization and modeling characteristics, extensibility and convergence characteristics, convertible and conductive characteristics, and integral and comprehensive characteristics.

1.2.1 Formalization and Modeling Characteristics

Contradictory problems in social science research are expressed with the natural human language. To enable people to deduce strategies for solving problems with certain procedures and generate strategies to solve problems by employing computers, Extenics adopts a formalized language to express matters, objects, relations, and problems. The problem extension model is established to present the process of quantitative changes, qualitative changes, and the critical state, as well as the process and tactics of generating strategies. Therefore the formal language can describe a process of solving contradictory problems, and it can be seen as an abstract model that reflects the inner relationship of the objects of concern by means of symbols.

1.2.2 Extensibility and Convergence Characteristics

An important characteristic of the Extenics methodology is that any object is extensible, while each extensible object converges under a certain condition. Conforming to people's thinking mode of "divergence→convergence" when solving contradictory problems, this is called the rhombus-thinking mode. A repeating process of "divergence→convergence→re-divergence→re-convergence" can be expressed by the multi-level rhombus-thinking mode. It stands for a thinking process, especially a formalized tool useful in a creative thinking process, as people's creative thinking process includes divergent searching and convergent synthesizing.

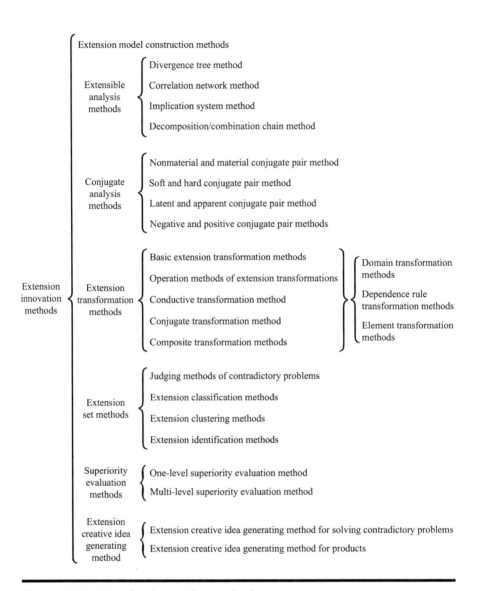

Figure 1.2.1 Extension innovation method system.

1.2.3 *Convertible and Conductive Characteristics*

Extenics studies qualitative and quantitative variability of things, as well as the transformative quality of "yes" and "no." It researches not only the formalization of direct transformation, but also the conduction mechanism of transformation. An important characteristic of the Extenics methodology is that it turns incompatible problems into compatible problems, it turns antithetical problems into coexisting problems, and it produces solutions for contradictory problems through formalized and quantified research tools.

1.2.4 Integral and Comprehensive Characteristics

Conjugate analysis is conducted on the things of concern as a whole from four angles by using a formalized model in Extenics. The conjugate analysis method, attempting to achieve comprehensive understanding of things, not only reflects the ancient Chinese ideas of systemic perspective and holism, but also combines the analysis method of reductionism. The organic combination of quality and quantity is reflected in the basic-element concept. Things of concern can be also analyzed as a whole with comprehensively characteristic basic-elements. In the extension set, the variation of values of dependent functions can be used to express processes of quantitative and qualitative changes, while the idea of solving contradictory problems from a holistic perspective is reflected in the domain transformation.

1.3 Introduction to Extension Innovation Method

1.3.1 Extension Model Construction Method

By adopting formalized languages to express things, objects, relations, and problems, a formal model, called an extension model, with basic-elements (i.e. objects, characteristics, values) as logic cells is established to enable people to deduce new product ideas or problem-solving strategies in accordance with certain procedures and to generate new product ideas or strategies to solve contradictory problems through using computers. It is an abstract model that reflects the inner relationship of the objects of concern by means of symbols.

The method of establishing a problem extension model consists of the following steps: (1) express the research object in the form of a basic-element or compound-element; (2) define the goal and condition of the problem; (3) describe the goal and condition of the problem with the basic-element or compound-element and abstract the core issue from the original problem; and (4) establish a function describing the contradictory degree of the problem for judging the contradictory degree of the issue.

To make the readers' experience enjoyable, only the construction method of the extension model for a research object is introduced in Chapter 2, while the construction method of the extension model for a problem is introduced in the Chapter 7.

1.3.2 Extensible Analysis Method

The process of innovation is also a process of solving all kinds of contradictory problems. Multiple solutions to contradictory problems can only be found by considering the problem-related object as one that can be extended. The extensible analysis method to express the extension rule of things can be established with basic-elements as formalized tools to describe objects, things, and relations to provide the formalization and modeling in the solution process of contradictory problems. This method not only can enable people to eliminate constraints in habitual domains, but it also can help people to solve contradictory problems and improve machine intelligence with the use of computers. The extensible analysis method includes the divergence tree method, the correlative network method, the implication system method, and the decomposition/combination chain method.

1.3.3 Conjugate Analysis Method

Whether it is a product innovation, a technological innovation, or an organizational innovation, the innovation is inseparable from the analysis of things. Different innovation schemes can be obtained by analyzing things from different perspectives. The conjugate analysis method established in Extenics involves analyzing objects from four aspects: materiality, systematicness, dynamism, and antagonism (collectively referred to as conjugacy). Also known as the conjugate pair method, consists of the material and nonmaterial conjugate pair method, the soft and hard conjugate pair method, the latent and apparent conjugate pair method, and the negative and positive conjugate pair method. In other words, by adopting matter-element and relation-element as formalized tools, this method can be employed to conduct qualitative analysis on the object's "material part, non-material part, and the intermediary part," on the "soft part, hard part, and the intermediary part," on the "latent part, apparent part, and the intermediary part," and on the "negative part, positive part, and the intermediary part." A variety of strategies for solving contradictory problems can be obtained by analyzing each conjugate part of the object as well as its mutual relationship and mutual transformation. Based on the idea of combining holism with reductionism, the conjugate analysis method provides a new perspective for people to comprehensively analyze things' structures, which is also a source of tactics for solving certain contradictory problems.

1.3.4 Extension Transformation Method

Extension transformation is an innovative tool. In the research of transformation, it is necessary not only to discuss the form of the transformation, but also to discuss the transformation of the discipline and the transformation method, tool, time, and place. That is, one must research both the transformation's form and connotation from qualitative and quantitative points of view. Direct transformations and indirect conductive transformation must be investigated. It is necessary to research not only the transformation of quantities, but also the transformation of characteristics and the object itself. Moreover, the form, connotation and conduction effect must be researched on the basis of the correlation among research objects.

In terms of the form of transformation, the extension transformation method includes the basic transformation method, the operation method of transformations, the composite method of transformations, and the conductive transformation method.

In terms of the object of transformation, the extension transformation method includes the transformation method of the universe of discourse, the transformation method of the relation maxim, and the transformation method of the elements in the domain. If the object of transformation is a thing, the extension transformation method also includes the transformation of the conjugate part and the conductive transformation of the conjugate part, which is referred to as the conjugate transformation method.

In terms of the composition of the extension model of contradictory problems, The extension transformation includes the transformations for goals and conditions of the problem.

Through the extension transformation method, an incompatible problem is transformed into a compatible problem; an antithetical problem is transformed into a coexisting problem; an unknowable problem is transformed into a known problem; an infeasible problem is transformed into a feasible problem; a false proposition is transformed into a true proposition; a false reasoning is transformed into a correct reasoning. These transformations are commonly referred to as ideas, tricks, and methods. Research on the extension transformation method provides operable tools for the formalization and quantification encountered in the process of solving contradictory problems together with the method of establishing dependent functions.

1.3.5 Extension Set Method

The extension set method is one developed to classify, identify, and cluster the research object from a dynamic and transforming perspective. Extension transformation and dependent function are two important components of the extension set. Against different extension transformations, the extension set has different domains of qualitative change and domains of quantitative change; thus, it also has different forms of classification, recognition, and clustering. It discloses the transformation process and the result of the contradictory problems of concern in a formalized and quantitative way, rendering classification, recognition, and clustering to be dynamic and convertible, which conforms to people's thinking mode and the actual situation under consideration.

The extension set method mainly includes the method of judging the given contradictory problem, the extension classification method, the extension clustering method, and the extension recognition method. It is a foundation of the extension data-mining method, which deals with large amounts of data in the database by using computers to obtain a basis of the extension knowledge. Because generating creative ideas is highlighted in this book, the extension set method will not be introduced individually. The required knowledge is introduced in Chapters 6 and 7. Readers who are interested in this part of contents can refer to the monograph ***Extenics*** for details.

1.3.6 Superiority Evaluation Method

The superiority evaluation method is a practical method developed for evaluating the advantages and disadvantages of a certain object, creative idea, strategy, etc., by integrating various measurement indexes. The conforming degree of each measurement index to the requirements is calculated with the dependent functions in the superiority evaluation method. Because the value of the dependent function can be either positive or negative, the established superiority can reflect the degree of advantages and disadvantages of an object, making the evaluation more realistic.

For a single measurement index, a simple dependent function, primary dependent function, discrete dependent function, interval-type dependent function, etc., can be selected according to the actual requirements of the measurement index. A comprehensive dependent function is established according to the requirements of the actual problem and professional knowledge, and this is used to calculate the comprehensive superiority of each object that is to be evaluated for judging its advantage and disadvantage or the grade of the object to be evaluated. The method that determines the weight coefficient can be selected appropriately according to the specifics of the problem.

The superiority evaluation method includes first-level superiority evaluation and multi-grade superiority evaluation. In the first-level superiority evaluation method, measurement indexes are not classified. The multi-level superiority evaluation method is applied in a situation with multiple measurement indexes, so these indexes should be classified first before endowing weights to all grades of the measurement indexes. After that, a comprehensive evaluation is conducted on the object to be evaluated.

1.3.7 Extension Creative Idea Generating Method

Generation of a creative idea represents a creative process of thinking, which is described as a "rhombus-thinking mode." That is, "it is a mode of firstly divergence and then convergence." As for its diverging process, most people think it is difficult to grasp because there seems to be no rule to follow. In fact, various creative ideas can be formed with the utilization of extensible analysis, conjugate analysis, and extension transformation after defining the problem appropriately by means of formalization or even a computer. It is a feasible formalized method in the divergence process, which contributes greatly to the generation of creative ideas.

Extension creative idea generation mainly includes two categories. The extension creative idea generating method for solving contradictory problems focuses on the creative idea generating method for solving the incompatible problem of concern. The method is also called extension strategic generation. The creative idea generation method for solving an antithetical problem is also known as the converting bridge method.

The extension creative idea generation method for new products and new projects has a four-step method to promote applications: modeling, extension, transformation, and evaluation. With this method, one can tell where the idea is from and how to obtain and determine a satisfactory and feasible idea. In terms of product innovation, there are three creation methods for generating new product ideas (i.e. formalization, quantification, and process), which can be developed into product innovation systems to help product innovation personnel create new products.

According to the establishment of the extension innovation method system and the preliminary application practice, it is feasible to apply the extension innovation method in formalizing and quantifying the creative thinking process, and to intelligently process contradictory problems through in-depth research. Further improvement of the extension innovation method system will inevitably drive the development of thinking science, decision science, and intelligence science and will improve the scientificity and operability of these related disciplinary areas.

1.4 Overview of the Application and Promotion of Extension Innovation Method

After years of research, the gradually maturing extension innovation method has been widely applied in the fields of engineering technology, information science, and intelligence science, economics, and management; it has also played an important role in product innovation, technology innovation, management innovation, and organizational innovation. The extension innovation method is an effective method to solve contradictory problems in the innovation fields of formalized and quantitative research, which is the prerequisite for the extension innovation method to be implemented using computer software. With the development of extension logic research, there are many extension softwares available for multiple fields, such as extension strategy generating system software, extension data mining software, extension design software, etc. In addition, with the development of extension control and extension testing research, many scholars have applied the extension innovation method in developing hardware products, obtaining a number of related patents.

The extension innovation method stylizes the process of solving contradictory problems for humans; it provides a method to complete the process of "detecting a problem → establishing a model for the problem → analyzing the problem → generating a strategy for solving the problem" with a formalized model. It is also a prerequisite for the extension innovation method to carry out socialized promotion and application.

The publication of popular books, such as *Revolution of Creative Ideas*, *Extension Innovation Thinking and Training*, and *Playing A Card Without Following Rules*, as well as the posting of videos online, such as "Intellectual Revolution," "Extenics," "Extension Innovation Method," "Interviewing Extenics Expert Cai Wen," etc., have created good publicity and cognitive conditions for the socialization of Extenics. Multiple media sources, like *People's Daily*, *Guangming Daily*, *Economic Daily*, *Science Times*, ScienceNet, Xinhua Net, People Net, Phoenix Net, etc., have reported on Extenics. Several universities have cultivated graduates in the research of Extenics and have offered courses on Extenics and the Extenics Innovation Method. Several Extenics workshops, seminars, and extension innovation training camps have been conducted in mainland China, Hong Kong, and Taiwan. In addition, extension innovation training has been conducted for several enterprises. A batch of patents and software copyrights applying the extension innovation method have been obtained, resulting in a good application basis in a number of fields.

In recent years, the easy-to-learn "extension innovation four-step method," which features popularity, convenience, and efficiency in its use, has been established to promote the extension innovation method in business circles and elementary and secondary schools, all of which can help generate ideas with the help of computers.

Two sessions of the "Enterprise Extension Innovation Method Backbone Training Course" have been held in Shenzhen since 2012, with over a hundred people attended the preliminary training. Four sessions of "Extension Innovation Teachers Training Course" have been held in Guangzhou, Jinan, and Beijing, with more than 120 teachers trained in the preliminary training course since 2014. Multiple sessions of Extenics seminars, training courses, and workshops have been held successively in Guangzhou, Daqing, Shenzhen, Beijing, Nanchang, Dalian, Harbin, Ningbo, Zhongshan, Yantai, Zhuhai, and other places. In addition, the grade certification work for the teachers of the extension innovation method was started in 2015, and more than thirty people have been awarded level-1 and level-2 certificates. There are now teachers with certificates who are conducting training on the enterprise extension innovation method and the youth extension innovation method. Tangible results have been obtained in the socialization of the extension innovation method.

According to the establishment of the extension innovation method system and the preliminary application practice, further improvement of the system will provide formal process and operational methods for innovation activities in various fields. That has an important application value for formalized quantitative research of technological innovation, invention, creation, etc.

The extension innovation method is an innovation method proposed by Chinese scientists independently. Although the time for its communication and application is short, it has already revealed its application value and has played an important role in practice. Promoting the extension innovation method is conducive to improving the quality of national innovation and helping innovators in various fields to achieve more innovative achievements.

Thinking and Exercises

1. Where is the scholar who founded Extenics from? What kind of discipline is Extenics in? What is the definition of Extenics?
2. What is the research object of Extenics? What is the basic theory? What is the method system?
3. What can an extension innovation method do?

Chapter 2

Starting Points of Innovation
Extension Model Construction Methods

Content Summary

Finding the starting point in an innovation process is the first step for innovation. Innovation is inseparable from the analysis of matters, affairs, and relations. Extenics establishes formalized basic-elements to describe matters, affairs and relations: matter-element, affair-element, and relation-element, which are collectively called basic-element. In addition, these basic-elements can form compound-elements to represent complicated matters and affairs. On such a basis, an extension model developed for a contradictory problem can be studied.

The purposes of formalization—standardization, refinement, quantification and computerization—are given.

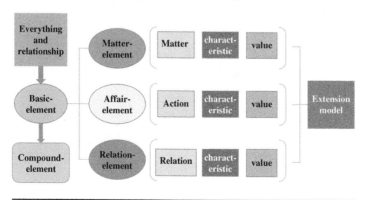

2.1 Modeling Representation of the Product—Matter-element

- Given a teacup, how many new cups can you think of?
- This is a product innovation problem—where do you get started?
- How many characteristics does the existing teacup have?
  You might first think about color, material, length, height, weight, etc. Which characteristics can you also think of? People with professional knowledge of cups may point out its more characteristics.
- How many colors can the cup have? How many kinds of materials can it have? How about its size?
- Changes in matters' names, characteristics, and values produce a variety of new things. Can you formalize, model, and standardize all of them? Is what you think of doing conducive to innovation?

2.1.1 *Definition of the Matter-element*

Definition 2.1 An ordered triplet constituted by matter O_m as the object, c_m as the characteristic, and v_m as the value of O_m about c_m, or $M = (O_m, c_m, v_m)$, is a basic element that describes the matter, called a one-dimensional matter-element. The symbols O_m, c_m, and v_m are three essential factors of the matter-element M. The two-tuple (c_m, v_m) constituted by c_m and v_m is the characteristic element of the matter O_m.

In product innovation, the matter-element is used as a tool for modeling products.

For example, $M_1 = (\text{Cup } D_1, \text{material, glass})$, where (material, glass) is the characteristic element of the one-dimensional matter-element M_1.

$$M_2 = (\text{Mobile phone } D_2, \text{weight}, 300\,\text{g})$$

$$M_3 = (\text{Chinese national flag } D_3, \text{color, red})$$

Any matter has multiple characteristics. Like a one-dimensional matter-element, multi-dimensional matter-elements can be defined:

Definition 2.2 The array constituted by matter O_m, n characteristic names c_{m1}, $c_{m2}, \cdots, c_{mn}$, and the corresponding value v_{mi} $(i = 1, 2, \cdots, n)$ of O_m about $c_{mi}(i = 1, 2, \cdots, n)$

$$M = \begin{bmatrix} O_m, & c_{m1}, & v_{m1} \\ & c_{m2}, & v_{m2} \\ & \vdots & \vdots \\ & c_{mn}, & v_{mn} \end{bmatrix} = \left(O_m, C_m, V_m \right)$$

is called an n-dimensional matter-element, where

$$C_m = \begin{bmatrix} c_{m1} \\ c_{m2} \\ \vdots \\ c_{mn} \end{bmatrix} \text{ and } V_m = \begin{bmatrix} v_{m1} \\ v_{m2} \\ \vdots \\ v_{mn} \end{bmatrix}.$$

For example,

$$M'_1 = \begin{bmatrix} \text{Teacup } D_1, & \text{material,} & \text{glass} \\ & \text{color,} & \text{red} \\ & \text{shape,} & \text{cylinder} \\ & \text{height,} & \text{10 cm} \end{bmatrix}$$

$$M'_2 = \begin{bmatrix} \text{Mobile phone } D_2, & \text{weight,} & \text{300 g} \\ & \text{size,} & \text{4.5 inch} \\ & \text{brand,} & \text{Samsung} \end{bmatrix}$$

$$M'_3 = \begin{bmatrix} \text{Chinese national flag } D_3, & \text{color,} & \text{red} \\ & \text{shape,} & \text{rectangle} \\ & \text{picture,} & \text{five stars} \\ & \text{weight,} & \text{100 g} \end{bmatrix}$$

Matters change with time t. For this reason, the concept of dynamic matter-elements is defined below.

Definition 2.3 In a matter-element $M = (O_m, c_m, v_m)$, if O_m and v_m are a function of time t, we call M a dynamic matter-element, denoted as

$$M(t) = \left(O_m(t), c_m, v_m(t) \right).$$

Then, $v_m(t) = c_m[O_m(t)]$. For convenient writing, we omit the parameter t and use $v_m = c_m(O_m)$ for short when there is no confusion. It describes the relationship between the matter and its value about some characteristic.

For multiple characteristics, we have a multi-dimensional dynamic matter-element, denoted as

$$
M(t) = \begin{bmatrix} O_m(t), & c_{m1}, & v_{m1}(t) \\ & c_{m2}, & v_{m2}(t) \\ & \vdots & \vdots \\ & c_{mn}, & v_{mn}(t) \end{bmatrix} = (O_m(t), C_m, V_m(t)).
$$

For example, the values of someone's age, height, and weight change with time t. This can be expressed by the following three-dimensional dynamic matter-element:

$$
M_4(t) = \begin{bmatrix} \text{Person } D_4(t), & \text{age}, & v_1(t) \\ & \text{height}, & v_2(t) \\ & \text{weight}, & v_3(t) \end{bmatrix}.
$$

Given a matter, it has a corresponding value about each characteristic, which at the same time is unique. When the value does not exist, use empty value Ø to represent it. If the value of O_m about characteristic c_m is not null, c_m is called a nonempty characteristic of O_m.

Definition 2.4 The matter-element corresponding to all nonempty characteristics of matter O_m,

$$
\begin{bmatrix} O_m, & c_{m1}, & v_{m1} \\ & c_{m2}, & v_{m2} \\ & \vdots & \vdots \\ & c_{mn}, & v_{mn} \\ & \vdots & \vdots \end{bmatrix},
$$

is called matter O_m's whole characteristic matter-element, denoted as cp $M(O_m)$.

At a fixed point in time, the whole characteristic matter-element of O_m is unique. For any two different matters O_{m1} and O_{m2}, at least one characteristic c_m can be found, satisfying $c_m(O_{m1}) \neq c_m(O_{m2})$.

For two matters $M_1 = (O_{m1}, c_{m1}, v_{m1})$, $M_2 = (O_{m2}, c_{m2}, v_{m2})$, if and only if $O_{m1} = O_{m2}, c_{m1} = c_{m2}, v_{m1} = v_{m2}$, M_1 and M_2 are equivalent, denoted as $M_1 = M_2$.

2.1.2 Elements of Matter-element

The fundamental factors in a matter-element include matter, characteristic, characteristic-element, and value. To extract these essential factors from real-life problems accurately, we must define clearly the classification of matters, the classification of characteristics, and the classification of values.

2.1.2.1 Individual Matter and Matter Class

There are all kinds of matters in the objective world. They all have many characteristics. Because of the differences in their characteristics, matters form different classes. The matters of which the values are in a certain range (or the definite value) form a class, and others do not belong to the class.

According to the denotation of matters, they can be divided into categorical matters and individual matters. For example, lamps and desk lamps are categorical matters, and a desk lamp is a subclass of the lamp. Desk lamp D is a specific desk lamp, an individual matter. In general, the values of an individual matter and a categorical matter about a same characteristic are different. For example, the weight of desk lamp D is 1 kg, while the desk lamp's value about the weight is an interval, such as <0, 4> kg. The value about the color is a discrete set.

Using the matter-element, the previously mentioned categorical matter and individual matter can be expressed clearly as the follows:

$$M_1 = \begin{bmatrix} \text{Desk lamp } D, & \text{weight,} & 1\,\text{kg} \\ & \text{color,} & \text{white} \end{bmatrix}$$

$$\{M\} = \begin{bmatrix} \{\text{Desk lamp}\}, & \text{weight,} & <0,4>\text{kg} \\ & \text{color,} & \{\text{white, red, ..., blue}\} \end{bmatrix}$$

This matter-element, which gives a formalized description of the categorical matter, is called a categorical matter-element.

Explanation 1 A matter is often made up of many parts. Therefore, when writing a matter-element, we should first list the characteristics and corresponding values of the whole matter, then decompose the matter into parts and write the matter-elements of the parts.

For example, a lamp consists of a lampshade, a lamp base, and a light rod. In the innovation of the desk lamp, in addition to using a matter-element to express the whole desk lamp, we should also use matter-elements to express its parts, which is more helpful to the planned innovation.

2.1.2.2 Characteristics and Characteristic-elements

A characteristic is an abstract result of an object's or a group of objects' characters. Characteristics are used to describe concepts. Any object or a set of objects has many characters. People abstract a concept according to the objects' common characters, and then the concept becomes the characteristic. Different professional fields focus on different characters of the same object. In a professional field, the characteristics that reflect the fundamental characters of the object under consideration are called essential characteristics. Therefore, the essential characteristics are different due to the different professional fields to which the concepts belong, reflecting various emphases of different professional fields.

To understand the matter of concern, it is critical for one to understand its characteristics and corresponding values (characteristic elements). The characteristic element is actually the "characteristic" that we often say in spoken language. For example, we often say, "a person's height is 1.8 m" in spoken language. That is actually a characteristic-element (height, 1.8 m). Some matter's "conductivity is very good." This expression is a characteristic element (conductivity, very good).

Explanation 2 The attributes, parameters, and factors described in some fields are classified as characteristics or characteristic-elements in Extenics. Some require the use of matter-elements to be expressed clearly.

For example, 39 engineering parameters are summarized in TRIZ. Taking the "weight of a moving object" and the "weight of a stationary object" as an example, in Extenics these parameters involves two characteristics: weight and motion state. The "motion state" is usually regarded as a parametric variable. So we have the following parametric matter-element

$$M(t) = \big(\text{Matter}\,D(t),\ \text{weight},\ v(t)\big)$$

Obviously, when $t =$ movement and $t =$ static, the value $v(t)$ may be different, or the same.

Explanation 3 Matter's components are not characteristics.

2.1.2.3 Value and Range of Values

The quantity, degree, or range of a matter about a characteristic is known as the value about this characteristic. The value can be quantitative or non-quantitative.

A value represented by a real number and a dimension is called a quantitative value. A value that is not represented by a real number is called a non-quantitative value. Non-quantitative values can be quantified, such as scoring, assigning, etc., for quantitative calculation.

Range Given a characteristic c_m, the range of its values is called the range of c_m, denoted as $V(c_m)$. For example, $V(\text{length}) = (0, +\infty)$, $V(\text{temperature}) = (-273°C, +\infty)$.

Value range The range of a matter O_m about characteristic c_m is known as the value range, denoted as $V_0(c_m)$. Obviously, $V_0(c_m) \subseteq V(c_m)$. For example, the value range of the class matter "Table" about the length is <0, 8> m, namely, V_0 (length) = <0, 8>m.

Note: In Extenics, an interval is expressed as *<a, b>*. It can represent an open interval, a closed interval, or a half-open, half-closed interval, which is different from the expression of intervals in classical mathematics.

2.2 Modeling Representation of the Product's Function—Affair-element

QUESTIONS AND THINKING:

- What is a teacup used for?
  To hold water, because this entity has a characteristic of volume.
- What else can it do?
  Press the paper, because it has the characteristic of weight.
  Decorate, because it has the characteristics of color and pattern.
  Health care, because of the special material it is made of.
- The essence of "function and use" of the product are all affairs!
- If the expressions of "function and use" can be formalized and standardized, are they conducive to innovation?

An interaction between one matter and another matter is called an affair. An affair is formally described by an affair-element. In product innovation, an affair-element is mainly used to describe the product's function and the user's needs.

Definition 2.5 Take an ordered triple constituted by an action O_a, a characteristic c_a, and a value of O_a' about c_a

$$A = (O_a, c_a, v_a)$$

as the basic element describing the affair, called a one-dimensional affair-element.

For example, A_1 = (Contain, dominant object, water D_1), A_2 = (Sell, dominant object, desk lamp D_2), A_3 = (Produce, dominant object, antique lamp D_3), ..., are all one-dimensional affair-elements. Each of these one-dimensional affair-elements expresses one incomplete event. To fully express an affair, we must consider other characteristics of the action.

The basic characteristics of an action are dominating object, acting object, receiving object, time, location, degree, mode, tool, and so on.

Definition 2.6 An array composed of an action O_a, n characteristics $c_{a1}, c_{a2}, ..., c_{an}$, and values of O_a' $v_{a1}, v_{a2}, ..., v_{an}$ about $c_{a1}, c_{a2}, ..., c_{an}$,

$$\begin{bmatrix} O_a, & c_{a1}, & v_{a1} \\ & c_{a2}, & v_{a2} \\ & \vdots & \vdots \\ & c_{an}, & v_{an} \end{bmatrix} = (O_a, C_a, V_a) \overset{\Delta}{=} A,$$

is called an *n*-dimensional affair-element, where

$$C_a = \begin{bmatrix} c_{a1} \\ c_{a2} \\ \vdots \\ c_{an} \end{bmatrix}, \quad V_a = \begin{bmatrix} v_{a1} \\ v_{a2} \\ \vdots \\ v_{an} \end{bmatrix}.$$

Each affair-element is a formalized expression of what to do, who will do it, for whom to do it, what time to do it, where to do it, the degree of doing it, how to do it, the tools to use, and so on.

For example:

$$A_1' = \begin{bmatrix} \text{Add,} & \text{dominating object,} & \text{water } D_1 \\ & \text{acting object,} & \text{young people } S_1 \\ & \text{receiving object,} & \text{parents } S_2 \\ & \text{tool,} & \text{teacup } S_3 \\ & \text{time,} & \text{2015 new year} \\ & \text{place,} & \text{at home} \end{bmatrix}$$

indicates the affair that "Young people S_1 added water D_1 to teacups S_3 for parents S_2 on the 2015 New Year's Day at home."

$$A_2' = \begin{bmatrix} \text{Sell,} & \text{dominating object,} & \text{desk lamp } D_2 \\ & \text{acting object,} & \text{exclusive shop } D_3 \\ & \text{receiving object,} & \text{company } S_4 \\ & \text{time,} & \text{April 2014} \\ & \text{place,} & \text{Guangzhou} \\ & \text{mode,} & \text{monopoly} \end{bmatrix}$$

indicates "Exclusive shop D_3 had a monopoly in April 2014 in Guangzhou and sold table Lamp D_2 for company S_4."

$$A_3' = \begin{bmatrix} \text{Produce,} & \text{dominating object,} & \text{antique lamp } D_3 \\ & \text{acting object,} & \text{enterprise } D_4 \\ & \text{time,} & \text{2010} \\ & \text{place,} & \text{Zhongshan} \\ & \text{mode,} & \text{by hand} \end{bmatrix}$$

indicates "Enterprise D_4 in 2010 in Zhongshan produced antique lamp D_3 by hand."

Because the function of the product, the consumer's need, and the enterprise's goal are all affairs, they can also be formally described as affair-elements.

For example, the statement that the goal of an enterprise D_4 is to "increase the market share by 10% in one year" can be formally expressed as an affair-element as follows:

$$A_4 = \begin{bmatrix} \text{Increase,} & \text{dominating object,} & \text{market share} \\ & \text{acting object,} & \text{enterprise } D_4 \\ & \text{time,} & \text{1 year} \\ & \text{degree,} & \text{10\%} \end{bmatrix}.$$

The statement that a lamp D_5 has the function of "providing light for learners S_5" can be formally expressed as the following affair-element:

$$A_5 = \begin{bmatrix} \text{Provide,} & \text{dominating object,} & \text{light} \\ & \text{tool,} & \text{desk lamp } D_5 \\ & \text{receiving object,} & \text{learner } S_5 \end{bmatrix}.$$

For the statement that "northern consumer S_6 needs something to protect her feet in spring," this "need" can be formally expressed by an affair-element as follows:

$$A_6 = \begin{bmatrix} \text{Protect,} & \text{dominating object,} & \text{feet} \\ & \text{acting object,} & \text{consumer } S_6 \\ & \text{time,} & \text{spring} \\ & \text{place,} & \text{north} \end{bmatrix}.$$

Some actions also have a direction, track, and other characteristics. For example, "Slide the mobile phone screen S_7 upward along a straight line" can be expressed by an affair-element as follows:

$$A_7 = \begin{bmatrix} \text{Slide,} & \text{dominating object,} & \text{mobile phone screen } S_7 \\ & \text{direction,} & \text{upward} \\ & \text{track,} & \text{straight line} \end{bmatrix}.$$

Definition 2.7 If in $A = (O_a, c_a, v_a)$, O_a and v_a are functions of time t, then A is said to be a dynamic affair-element, denoted as

$$A(t) = \big(O_a(t), c_a, v_a(t)\big).$$

For a multi-dimensional affair-element, we have

$$A(t) = \big(O_a(t), C_a, V_a(t)\big).$$

For example, with the change of time t, the values of the action "teaching" about dominating object, acting object, place, and way may change, which can be expressed by an dynamic affair-element as follows:

$$
A(t) = \begin{bmatrix}
\text{Teach}(t), & \text{dominating object,} & v_{a1}(t) \\
& \text{acting object,} & v_{a2}(t) \\
& \text{place,} & v_{a3}(t) \\
& \text{mode,} & v_{a4}(t)
\end{bmatrix}.
$$

Likewise, with the change of time t, consumers may have different requirements for the tools and methods of "decorating room D_1," which can be expressed by a dynamic multi-dimensional affair-element as follows:

$$
A(t) = \begin{bmatrix}
\text{Decorate}(t), & \text{dominating object,} & \text{room } D_1(t) \\
& \text{acting object,} & \text{young people } S_1 \\
& \text{receiving object,} & \text{parents } S_2 \\
& \text{tool,} & \text{light } (t) \\
& \text{mode,} & \text{flashing } (t) \\
& \text{place,} & \text{at home}
\end{bmatrix}.
$$

Similar to the concept of matter-elements, a formal representation of a class affair is called a class affair-element. The values of a class affair-element about some characteristics of a verb may be class values. For example:

$$
\{A\} = \begin{bmatrix}
\text{Decorate,} & \text{dominating object,} & \{\text{room, hall, kitchen}\} \\
& \text{acting object,} & \{\text{young people, aged people, children}\}
\end{bmatrix}.
$$

The equality of affair-elements is such that given affair-elements $A_1 = (O_{a1}, c_{a1}, v_{a1})$ and $A_2 = (O_{a2}, c_{a2}, v_{a2})$, $A_1 = A_2$ if and only if $O_{a1} = O_{a2}, c_{a1} = c_{a2}, v_{a1} = v_{a2}$.

2.3 Modeling Representation of the Product's Structure—Relation-element

QUESTIONS AND THINKING:

■ Many innovative idea generations or resolutions of contradiction problems can be accomplished by changing relationships. For example: change the "upper and lower relationship" between "lamp shade" and "lamp holder" from the original "the lamp shade on the top and the lamp holder at the bottom" into "the lamp holder on the top and the lamp shade on the bottom." Can the result become a new type of lamp? Can changing the "control relationship" of the lamp create a new type of lamp?

- How can these "relationships" be thought out and analyzed in a more orderly way?
- If the expression of "relationship" can be formalized and standardized, is it conducive to innovation?

In the real world, there are countless relationships among matters, affairs, people, information, and knowledge. As these relationships interact and influence each other, there are also various relationships among the corresponding matter-elements and affair-elements. Changes in these relationships will also interact and influence each other. Relation-element is a formal tool to describe such phenomena.

In product innovation, the concept of relation-elements is mainly used for modeling products' structure relations.

Definition 2.8 The following n-dimension array composed of relation words or relation character (relation name for short) O_r, n characteristics $c_{r1}, c_{r2},..., c_{rn}$, and corresponding values $v_{r1}, v_{r2},..., v_{rn}$:

$$\begin{bmatrix} O_r, & c_{r1}, & v_{r1} \\ & c_{r2}, & v_{r2} \\ & \vdots & \vdots \\ & c_{rn}, & v_{rn} \end{bmatrix} = (O_r, C_r, V_r) \overset{\Delta}{=} R$$

is used to describe the relation between v_{r1} and v_{r2}, called an n-dimensional relation-element, where

$$C_r = \begin{bmatrix} c_{r1} \\ c_{r2} \\ \vdots \\ c_{rn} \end{bmatrix}, \quad V_r = \begin{bmatrix} v_{r1} \\ v_{r2} \\ \vdots \\ v_{rn} \end{bmatrix}.$$

For example,

$$R_1 = \begin{bmatrix} \text{Connection relation,} & \text{antecedent,} & \text{lamp holder } D_1 \\ & \text{consequent,} & \text{bulb } D_2 \\ & \text{degree,} & 100 \\ & \text{maintain way,} & \text{embed} \end{bmatrix}$$

$$R_2 = \begin{bmatrix} \text{Up and down relation,} & \text{antecedent,} & \text{lampholder } D_1 \\ & \text{consequent,} & \text{bulb } D_2 \\ & \text{degree,} & 100 \end{bmatrix}$$

describe the connection relation and the up and down position relation between the lamp holder and the bulb.

And the control relation of switch D_1 and lamp D_2 can be formally described as the following multi-dimensional relation-element:

$$R_3 = \begin{bmatrix} \text{Control relation,} & \text{antecedent,} & \text{switch } D_1 \\ & \text{consequent,} & \text{lamp } D_2 \\ & \text{degree,} & \text{close} \\ & \text{maintain way,} & \text{press} \\ & \text{contact channel,} & \text{electric wire} \\ & \text{contact way,} & \text{electricity} \\ & \text{place,} & D \text{ place} \end{bmatrix}$$

Among these previously described characteristics, the former term, the latter term, and degree are common characteristics. They express the objects and degree of the relationship.

In product innovation, sometimes it is possible to produce a new product by changing the relation name or by changing the value of a characteristic in the relation-element. For example, changing the "up and down relation" to "left-right relation," the "embed" to "spiral into," and so on, can produce new products.

Definition 2.9 In a relation-element R, if the relation described by R is function of time t, we call

$$R(t) = \begin{bmatrix} O_r(t), & c_{r1}, & v_{r1}(t) \\ & c_{r2}, & v_{r2}(t) \\ & \vdots & \vdots \\ & c_{rn}, & v_{rn}(t) \\ & \vdots & \vdots \end{bmatrix}$$

a dynamic relation-element. $R(t)$ describes the dynamic change (including the change of the relation degree) of the relation O_r between v_{r1} and v_{r2} caused by the change of time t. The influence of different people, affairs, and matters also makes the relation change. These changes represent a change of the relation degree. A change in relation degree indicates relations' establishing, deepening, interrupting, worsening, and so on. It can be positive, zero, or negative in value.

For the following two relation-elements,

$$R_1 = \begin{bmatrix} O_{r1}, & c_{r1}, & v_{r11} \\ & c_{r2}, & v_{r12} \\ & \vdots & \vdots \\ & c_{rm}, & v_{r1n} \end{bmatrix}, \quad R_2 = \begin{bmatrix} O_{r2}, & c_{r1}, & v_{r21} \\ & c_{r2}, & v_{r22} \\ & \vdots & \vdots \\ & c_{rm}, & v_{r2n} \end{bmatrix},$$

if $O_{r1} = O_{r2}$, and for all $i \in \{1, 2, \ldots, n\}$, $v_{r1i} = v_{r2i}$, then we say that these two relation-elements are equal, denoted as $R_1 = R_2$.

When solving contradictory problems, people have to deal with a large number of people, affairs, matters, and relations. One of the basic tasks of the decision maker is to clarify all relationships among people, matters and affairs, based on which creative thinking is carried out to produce satisfactory resolutions. Therefore, being able to recognize these fundamental relations is particularly important. Actually, knowing these relations in essence is a challenging exploration process that requires eliminating that which is false and keeping that which is true, and moving from being coarse to being fine, from outward appearance to inner essence.

Brief Summary

Such basics as matter-elements, affair-elements, and relation-elements are jointly called basic-elements. To avoid causing confusion, we denote a basic-element as

$$B = \begin{bmatrix} \text{Object}, & c_1, & v_1 \\ & c_2, & v_2 \\ & \vdots & \vdots \\ & c_n, & v_n \end{bmatrix}$$

where O (Object) represents an object (matter, action, or relation name), $c_1, c_2, \ldots, c_n$ the n characteristics of the object O, and $v_1, v_2, \ldots, v_n$ the corresponding values of object O's individual characteristics.

If a basic-element B is a function of time t, we have the following dynamic basic-element

$$B(t) = \begin{bmatrix} \text{Object}(t), & c_1, & v_1(t) \\ & c_2, & v_2(t) \\ & \vdots & \vdots \\ & c_n, & v_n(t) \end{bmatrix}.$$

The class basic-element is represented as

$$\{B\} = \begin{bmatrix} \{\text{Object}\}, & c_1, & V_1 \\ & c_2, & V_2 \\ & \vdots & \vdots \\ & c_n, & V_n \end{bmatrix},$$

where $V_i (i = 1, 2, \cdots, n)$ is the value range of the class object $\{Object\}$ about characteristic c_i.

2.4 Modeling Representation of Complex Matters, Affairs, and Relations

QUESTIONS AND THINKING:

■ The value of affair and relation about a characteristic may be a matter, which in turn has its own characteristic and value. This means that merely using matter-elements, affair-elements, and relation-elements is not enough to clearly express such complex structures.
■ How can one formally express the control relation between a switch and multiple lights or multiple switches and one light?
■ How can one formally express "decorate a 50-m² room using colorful lights"?
■ Is it conducive to innovation if one can express complex affairs, matters, and relationships in a formalized and standardized way?

2.4.1 Introduction to Compound-elements

Problems in the real world are often very complex. They tend to be combinations or compounded results of people, affairs, and matters. We therefore need to use the compound form of matter-elements, affair-elements, and relation-elements to express the objects of these problems. These compound forms are uniformly called compound-elements. Studying the composition, operation, and transformation of compound-elements constitutes the basis of studying complex problems.

A compound-element can take many different forms. Here are just a few commonly used forms. Interested readers can refer to the relevant content in the monograph **Extenics**.

For example, the statement "decorate the 50-m² room with colorful lights" can be expressed as a compound-element consisting of a matter-element and an affair-element as follows:

$$
A(M) = \begin{bmatrix}
\text{Decorate,} & \text{dominating object,} & (\text{Room } D_1, \text{ area, } 50 \text{ m}^2) \\
 & \text{tool,} & (\text{Light } D_2, \text{ color, } 7 \text{ color}) \\
 & \text{place,} & \text{at home}
\end{bmatrix}.
$$

As another example, the phrase "control relationship between a switch and six bulbs" can be expressed using a compound-element consisting of a matter-element and a relation element:

$$
R_1(M_1, M_2) = \begin{bmatrix}
\text{Control relation,} & \text{antecedent,} & (\text{Switch } D_1, \text{ number, } 1) \\
 & \text{consequent,} & (\text{Bulb } D_2, \text{ number, } 6)
\end{bmatrix}.
$$

The phrase "control relationship between two switches and one bulb" can be expressed using a compound-element compounded consisting of a matter-element and a relation element:

$$R_2(M_3, M_4) = \begin{bmatrix} \text{Control relation,} & \text{antecedent,} & (\text{Switch } D_3, \text{ number, 2}) \\ & \text{consequent,} & (\text{Bulb } D_4, \text{ number, 1}) \end{bmatrix}$$

If one wants to express "control six bulbs with a switch: press the switch one time to light up two bulbs; press the switch two times to light up four bulbs; press the switch three times to light up six bulbs; press the switch four times to turn off all bulbs," the above compound-elements can be used. The operations of compound-elements or transformations are required for a clear expression. Related details are omitted here.

2.4.2 Logical Operations of Basic-elements

To describe complex affairs, matters, and relationships, in addition to the application of matter-elements, affair-elements, relations elements, and compound-elements, some operations between basic-elements and compound-elements are also required. The following is a brief introduction to commonly used logical operations among basic-elements, including the AND operation, the OR operation, and the NOT operation. These operations among compound-elements are more complicated, so we do not introduce them here.

2.4.2.1 AND Operation of Basic-elements

Given basic-element $B_1 = (O_1, c_1, v_1)$, $B_2 = (O_2, c_2, v_2)$, the "AND operation" of B_1 and B_2 is to take both of them, denoted as

$$B = B_1 \wedge B_2 = (O_1 \wedge O_2, \quad c_1 \wedge c_2, \quad v_1 \wedge v_2)$$

$$= \begin{cases} (O, \quad c, \quad v_1 \wedge v_2), & \text{if } O_1 = O_2 = O, c_1 = c_2 = c \\ (O_1 \wedge O_2, \quad c, \quad v_1 \wedge v_2), & \text{if } O_1 \neq O_2, \quad c_1 = c_2 = c \\ \begin{bmatrix} O, & c_1, & v_1 \\ & c_2, & v_2 \end{bmatrix}, & \text{if } O_1 = O_2 = O, c_1 \neq c_2 \\ \begin{bmatrix} O_1 \wedge O_2, & c_1, & v_1 \wedge v_{21} \\ & c_2, & v_{12} \wedge v_2 \end{bmatrix}, & \text{if } O_1 \neq O_2, \quad c_1 \neq c_2 \end{cases}$$

where v_{21} is the value of object O_2 about characteristic c_1, v_{12} is the value of object O_1 about characteristic c_2.

2.4.2.2 OR Operation of Basic-Elements

Given basic-element $B_1 = (O_1, c_1, v_1)$, $B_2 = (O_2, c_2, v_2)$, the "OR operation" of B_1 and B_2 is to take at least one of them, denoted as

$$B = B_1 \vee B_2 = (O_1 \vee O_2, \quad c_1 \vee c_2, \quad v_1 \vee v_2)$$

$$= \begin{cases} (O, \quad c, \quad v_1 \vee v_2), & \text{if } O_1 = O_2 = O, c_1 = c_2 = c \\ (O_1 \vee O_2, \quad c, \quad v_1 \vee v_2), & \text{if } O_1 \neq O_2, \quad c_1 = c_2 = c \\ (O, \quad c_1 \vee c_2, \quad v_1 \vee v_2), & \text{if } O_1 = O_2 = O, c_1 \neq c_2 \\ \begin{bmatrix} O_1 \vee O_2, & c_1, & v_1 \vee v_{21} \\ & c_2, & v_{12} \vee v_2 \end{bmatrix}, & \text{if } O_1 \neq O_2, \quad c_1 \neq c_2 \end{cases}$$

Example 2.4.1

Suppose $M_1 = (\text{Desk } D_1, \text{length}, 1 \text{ m})$, $M_2 = (\text{Chair } D_2, \text{length}, 0.5 \text{ m})$, then

$$M_1 \wedge M_2 = (\text{Desk } D_1 \wedge \text{Chair } D_2, \text{ length}, 1 \text{ m} \wedge 0.5 \text{ m})$$

means at the same time taking the matter-element M_1 and M_2, while

$$M_1 \vee M_2 = (\text{Desk } D_1 \vee \text{Chair } D_2, \text{ length}, 1 \text{ m} \vee 0.5 \text{ m})$$

represents taking at least one of the matter-elements M_1 and M_2.

Example 2.4.2

If the two functional affair-elements below are given,

$$A_1 = \begin{bmatrix} \text{Provide,} & \text{dominating object,} & \text{light} \\ & \text{tool,} & \text{desk lamp } D \\ & \text{place,} & \text{bedroom} \end{bmatrix},$$

$$A_2 = \begin{bmatrix} \text{Play,} & \text{dominating object,} & \text{music} \\ & \text{tool,} & \text{desk lamp } D \\ & \text{place,} & \text{bedroom} \end{bmatrix},$$

then $A_1 \wedge A_2$ means "desk lamp D provides both light and music in the bedroom," that is

$$A_1 \wedge A_2 = \begin{bmatrix} \text{Provide} \wedge \text{Play,} & \text{dominating object,} & \text{light} \wedge \text{music} \\ & \text{tool,} & \text{desk lamp } D \\ & \text{place,} & \text{bedroom} \end{bmatrix}$$

$A_1 \vee A_2$ means "desk lamp D provides light or plays music in the bedroom, that is

$$A_1 \vee A_2 = \begin{bmatrix} \text{Provide} \vee \text{Play}, & \text{dominating object}, & \text{light} \vee \text{music} \\ & \text{tool}, & \text{desk lamp } D \\ & \text{place}, & \text{bedroom} \end{bmatrix}$$

For basic-elements B_1 and B_2, obviously we have $B_1 \wedge B_2 = B_2 \wedge B_1$ and $B_1 \vee B_2 = B_2 \vee B_1$.

We can also define the AND operation and OR operation of multiple basic-elements. All details are omitted here.

2.4.2.3   NOT Operation of Basic-Elements

The NOT operation of basic-element $B = (O, c, v)$ includes "not the object" and "not the value," denoted respectively as

$$\overline{B}_O = (\overline{O}, c, v), \quad \overline{B}_v = (O, c, \overline{v}).$$

Example 2.4.3

If

$$M = \begin{bmatrix} \text{Air conditioner } D, & \text{service life}, & 10 \text{ years} \\ & \text{cost}, & 10,000 \text{ yuan} \end{bmatrix},$$

then

$$\overline{M}_O = \begin{bmatrix} \text{Air conditioners } \overline{D}, & \text{service life}, & 10 \text{ years} \\ & \text{cost}, & 10,000 \text{ yuan} \end{bmatrix}$$

means all air conditioners except D of which the service life is 10 years and the cost is 10,000 yuan, while

$$\overline{M}_v = \begin{bmatrix} \text{Air conditioners } D, & \text{service life}, & \overline{10} \text{ years} \\ & \text{cost}, & \overline{10,000} \text{ yuan} \end{bmatrix}$$

means air conditioners D of which "the service life is not 10 years and the cost is not 10,000 yuan."

Example 2.4.4

If

$$A = \begin{bmatrix} \text{Play}, & \text{dominating object}, & \text{music} \\ & \text{tool}, & \text{desk lamp } D \\ & \text{place}, & \text{bedroom} \end{bmatrix},$$

then

$$\bar{A}_O = \begin{bmatrix} \overline{\text{Play}}, & \text{dominating object}, & \text{music} \\ & \text{tool}, & \text{desk lamp } D \\ & \text{place}, & \text{bedroom} \end{bmatrix},$$

$$\bar{A}_v = \begin{bmatrix} \text{Play}, & \text{dominating object}, & \overline{\text{music}} \\ & \text{tool}, & \text{desk lamp } D \\ & \text{place}, & \text{bedroom} \end{bmatrix}$$

mean "the desk lamp in the bedroom D does not play music" and "the desk lamp in the bedroom D plays non-music," respectively.

Example 2.4.5

If $R = \begin{bmatrix} \text{upper and lower relation}, & \text{antecedent}, & D_1 \\ & \text{consequent}, & D_2 \end{bmatrix},$

then

$$\bar{R}_O = \begin{bmatrix} \overline{\text{upper and lower relation}}, & \text{antecedent}, & D_1 \\ & \text{consequent}, & D_2 \end{bmatrix} \text{ and}$$

$$\bar{R}_v = \begin{bmatrix} \text{upper and lower relation}, & \text{antecedent}, & \bar{D}_1 \\ & \text{consequent}, & \bar{D}_2 \end{bmatrix}$$

mean "Non-upper and lower relationship between D_1 and D_2," "upper and lower relationship between non-D_1 and non-D_2," respectively.

Explanation The NOT operation of basic-element B is expressed by symbol "$\neg B$." Using this symbol, the "not the object" and the "not the value" of the basic-element can be expressed as

$$\neg B_O = (\neg O, c, v), \quad \neg B_v = (O, c, -v).$$

Thinking and Exercises

1. Starting with a ceramic cup, how many characteristics and corresponding values of it can you write? Please use multi-dimensional matter-elements to express your answer. Can you think of three new products based on some of these characteristics?

2. Starting with a desk lamp, how many characteristics and corresponding values of it can you write? Please use multi-dimensional matter-elements to express your answer. Can you think of three new products based on some of these characteristics?

3. Please use multi-dimensional affair-elements to represent the function of a laser pen. Can you think of three products with new functions based on some of these characteristics?
4. Please use multi-dimensional affair-elements to represent the function of a desk lamp. Can you think of three products with new functions based on some of these characteristics?
5. Please use multi-dimensional relation-elements to describe the relationship between the pen cap and the penholder of a chosen pen. Can you think of three products with new relations based on this relationship?

Chapter 3

Basis for Generating Creative Ideas (1)
Extensible Analysis Methods

Content Summary

Extensible analysis extends basic-elements in a formalized way to obtain a variety of innovative paths or various ways for solving contradictory problems.

Extensible analysis methods include divergence analysis method, correlation analysis method, implication analysis method, and opening-up analysis method. According to the form of the results extended by these methods, they correspondingly lead to the divergence tree method, correlation network method, implication system (tree) method, and decomposition/combination chain method.

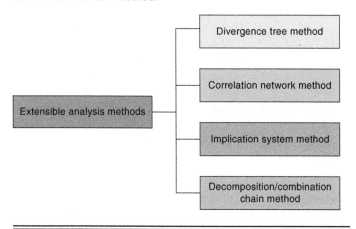

3.1 Divergence Tree Method

QUESTIONS AND THINKING:

■ Given a sheet of paper, what can it be used for? Why?
Because paper has a good "writing" ability, you can use it to write;
Because paper has a good "folding" ability, you can use it for folding aircraft;
Because the paper's "color" is red, you can use it to cut out a stop sign;
Because the paper has a good "absorbing" capability, you can use it to soak up the water on the table;
Because the paper has "folding" ability and "thickness", it can be used to prevent the uneven table or cabinet from tilting.
■ What else has the same functions? Why?

Take away: Any object has an extensible property. Different characteristics correspond to different uses.

Pre-knowledge: Divergence Rule

Divergence rule 3.1.1 One object-multiple characteristics-multiple values. Namely, from a given basic-element, we obtain a number of basic-elements with the same object, that is, different characteristics and values. This can be represented as follows:

$$B = \left(O, \quad c, \quad v \right) \dashv \left\{ \left(O, \quad c_1, \quad v_1 \right), \left(O, \quad c_2, \quad v_2 \right), ..., \left(O, \quad c_n, \quad v_n \right) \right\}$$
$$= \left\{ \left(O, \quad c_i, \quad v_i \right), i = 1, 2, ..., n \right\}$$

Depending on the definition of the multidimensional basic-element, this previous form can also be rewritten as

$$B = \left(O, \quad c, \quad v \right) \dashv \begin{bmatrix} O, & c & v \\ & c_1, & v_1 \\ & \vdots & \vdots \\ & c_n, & v_n \end{bmatrix}$$

According to this rule, when innovating or dealing with contradictory problems, if the existing basic-element doesn't work, one can consider using an alternative basic-element formed by the object of the existing basic-element and other characteristics.

This rule corresponds to the relational data Table 3.1.1.

Divergence rule 3.1.2 Multiple objects-one characteristic-multiple values. Namely, from a given basic-element, we can obtain a number of basic-elements

Table 3.1.1 The Relational Data Table of Divergence Rule 3.1.1

value \ characteristic \ object	c	c_1	c_2	...	c_n
O	v	v_1	v_2	...	v_n

with the same characteristics, that is, different objects and different values. This fact can be represented as follows:

$$B = (O, \ c, \ v) \dashv \{(O_1, \ c, \ v_1), \ (O_2, \ c, \ v_2), \ ..., \ (O_n, \ c, \ v_n)\}$$
$$= \{(O_i, \ c, \ v_i), \ i = 1, \ 2, \ ..., \ n\}$$

According to this rule, when innovating or dealing with contradictory problems, if the existing basic-element doesn't work, one can consider using an alternative basic-element formed by other objects and corresponding values about the same characteristics as the existing basic-element.

This rule can also be represented by the relational data Table 3.1.2.

Divergence rule 3.1.3 One object-multiple characteristics-one value. Namely, from a given basic-element, one can obtain a number of basic-elements with the same object, that is, different characteristics but the same value. This fact can be represented as follows:

$$B = (O, \ c, \ v) \dashv \{(O, \ c_1, \ v), \ (O, \ c_2, \ v) ..., \ (O, \ c_n, \ v)\}$$
$$= \{(O, \ c_i, \ v), \ i = 1, \ 2, \ ..., \ n\}$$

Table 3.1.2 The Relational Data Table of Divergence Rule 3.1.2

value \ characteristic \ object	c
O	v
O_1	v_1
O_2	v_2
⋮	⋮
O_n	v_n

Table 3.1.3 The Relational Data Table of Divergence Rule 3.1.3

characteristic value object	c	c_1	c_2	...	c_n
O	v	v	v	...	v

This rule corresponds to the relational data Table 3.1.3.

Divergence rule 3.1.4 Multiple objects-multiple characteristics-one value. Namely, from a given basic-element, one can obtain a number of basic-elements with different objects, that is, different characteristics but the same value. This fact can be represented as follows:

$$B = \left(O, \quad c, \quad v \right) \dashv \left\{ \left(O_1, \quad c_1, \quad v \right), \left(O_2, \quad c_2, \quad v \right), ..., \left(O_n, \quad c_n, \quad v \right) \right\}$$
$$= \left\{ \left(O_i, \quad c_i, \quad v \right), i = 1, 2, ..., n \right\}$$

This rule corresponds to the relational data Table 3.1.4.

Divergence rule 3.1.5 Multiple objects-one characteristic-one value. Namely, from a basic-element, one can obtain a number of basic-elements with different objects but the same characteristic and value. This rule can be represented as follows:

$$B = \left(O, \quad c, \quad v \right) \dashv \left\{ \left(O_1, \quad c, \quad v \right), \left(O_2, \quad c, \quad v \right), ..., \left(O_n, \quad c, \quad v \right) \right\}$$
$$= \left\{ \left(O_i, \quad c, \quad v \right), i = 1, 2, ..., n \right\}$$

Table 3.1.4 The Relational Data Table of Divergence Rule 3.1.4

characteristic value object	c	c_1	c_2	...	c_n
O	v				
O_1		v			
O_2			v		
⋮				...	
O_n					v

Table 3.1.5   The Relational Data Table of Divergence Rule 3.1.5

characteristic value object	c
O	v
O_1	v
O_2	v
$\vdots$	$\vdots$
O_n	v

This rule corresponds to the relational data Table 3.1.5.

Divergence rule 3.1.6 One object-one characteristic-multiple values. Namely, from a given parametric basic-element, one can expand to a number of parametric basic-elements with different values under different parametric variables but with the same characteristic. This rule can be represented as follows:

$$B(t) = \left(O(t), \quad c, \quad v(t)\right)$$
$$\dashv \left\{\left(O(t_1), \quad c, \quad v_1(t_1)\right), \ \left(O(t_2), \quad c, \quad v_2(t_2)\right), \ldots, \ \left(O(t_n), \quad c, \quad v_n(t_n)\right)\right\}$$
$$= \left\{\left(O(t_i), \quad c, \quad v_i(t_i)\right), \ i = 1, 2, \ldots, n\right\}$$

To avoid causing confusion, the parameter variable can be omitted.

This rule corresponds to the relational data Table 3.1.6.

Table 3.1.6   The Relational Data Table of Divergence Rule 3.1.6

characteristic value object	c
$O(t)$	$v(t)$
$O(t_1)$	$v_1(t_1)$
$O(t_2)$	$v_2(t_2)$
$\vdots$	$\vdots$
$O(t_n)$	$v_n(t_n)$

Basic Steps of the Divergent Tree Method

According to the divergence rules, starting from a basic-element, we can deduce several basic-elements, thus providing many possible paths to innovation or solving contradictory problems.

In the process of solving practical problems, we may find the superior paths to innovation or solving contradictory problems using only one divergent rule, and sometimes we must comprehensively apply several rules. Such a divergent process forms a tree-like structure and is thus called a divergence tree.

The general model of a basic-element divergence tree is as follows:

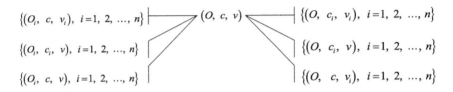

$$\{(O_i, c, v_i), \ i=1, 2, ..., n\} \quad \quad (O, c, v) \quad \quad \{(O, c_i, v_i), \ i=1, 2, ..., n\}$$
$$\{(O_i, c_i, v), \ i=1, 2, ..., n\} \quad \quad \{(O, c_i, v), \ i=1, 2, ..., n\}$$
$$\{(O_i, c, v), \ i=1, 2, ..., n\} \quad \quad \{(O, c, v_i), \ i=1, 2, ..., n\}$$

The method of using a divergence rule to find paths to innovation or a solution to a contradictory problem is called the divergent tree method. The basic steps of the method are given below:

1. List the basic-element B to be analyzed;
2. According to the given problem, optionally apply the divergence rules;
3. Deduce multiple basic-elements $(B_1, B_2, ..., B_n)$ from B;
4. Determine whether a path to innovation or a solution to the contradictory problem is found. If so, stop; otherwise, execute the next step;
5. Continue to apply the above procedure starting from B_i until you find a path to innovation or a solution to the contradictory problem.

Case Analysis:

Example 3.1.1

For a given wire D, everyone knows that it has electric conductivity, that is, we have the matter-element $M = (\text{Wire } D, \text{ conductivity}, v_m)$. Applying Divergence rule 3.1.1, we have

$$M = \left(\text{Wire } D, \text{ conductivity}, v_m\right) \dashv \left\{\left(\text{Wire } D, \text{ flexibility}, v_{m1}\right),\right.$$
$$\left.\left(\text{Wire } D, \text{ length}, v_{m2}\right), \left(\text{Wire } D, \text{ color}, v_{m3}\right), ...\right\}$$

When the wire is used to connect to an electrical source, we should consider its conductivity; when it is used to tie things together, we should consider its flexibility; when it is used for decoration or distinguished from other wires, we should consider its color.

For the wire D, of course, there are many other characteristics. However, please note that the deduction should be performed based on specific given problems. While considering the expansion of characteristics and the value of the basic-element, we should not artificially add inappropriate restrictions to the object. Otherwise, we will be unable to solve the problem—or we may even make a solvable problem unsolvable.

Example 3.1.2

The game of setting patterns that everyone often plays with matches is: Use six matchsticks to put create four regular triangles. This is a solvable problem, but many people think it is a contradictory problem. The main reason is the following improper expansion of the matter-element:

$$M = \left(\text{Triangle group } D_1, \quad \text{triangle number,} \quad 4\right)$$

$$\underline{\text{rule 3.1.1}} \left| \begin{cases} M_1 = \left(\text{Triangle group } D_1, \quad \text{location,} \quad \text{plane}\right) \\ M_2 = \left(\text{Triangle group } D_1, \quad \text{triangular edge length,} \quad a\right) \end{cases} \right.$$

where a is the length of each matchstick. That is to say, matter-elements M_1 and M_2 are not only irrelevant, but they also become an obstacle to possibly solving the problem because of their inappropriate values.

For different times t_1, t_2, and t_3, basic-elements M_1 and M_2 extended from M can be further extended according to rule 3.1.6:

$$M_1(t_1) = \left(\text{Triangle group } D_1(t_1), \quad \text{location,} \quad \text{plane}(t_1)\right)$$

$$\underline{\text{rule 3.1.6}} \Big| M_{11}(t_2) = \left(\text{Triangle group } D_1(t_2), \quad \text{location,} \quad \text{space}(t_2)\right)$$

$$M_2(t_1) = \left(\text{Triangle group } D_1(t_1), \quad \text{triangle edge length,} \quad a(t_1)\right)$$

$$\underline{\text{rule 3.1.6}} \Big| M_{21}(t_3) = \left(\text{Triangle group } D_1(t_3), \quad \text{triangle edge length,} \quad a'(t_3)\right)\left(a' < a\right)$$

According to the condition matter-element $l = (\text{match group } D_2, \text{ number, } 6)$ and the three-dimensional matter-element formed by M, M_{11}, M_2 at time t,

$$M'(t) = \begin{bmatrix} \text{Triangle group } D_1(t), & \text{triangle number,} & 4 \\ & \text{location,} & \text{space}(t) \\ & \text{triangle edge length,} & a(t) \end{bmatrix},$$

a three-dimensional graphic can be posed—a regular tetrahedron that has four positive triangles on it. Again, from the following three-dimensional matter-element formed by M, M_1, M_{21} at time t',

$$M''(t') = \begin{bmatrix} \text{Triangle group } D_1(t'), & \text{triangle number,} & 4 \\ & \text{location,} & \text{plane}(t') \\ & \text{triangle edge length,} & a'(t') \end{bmatrix},$$

you can create many graphics whose side lengths are less than the match branch length a on the plane. Each graphic has four regular triangles on it.

Example 3.1.3

A teacup D_1 has the characteristic element (volume, 500 mL). A bowl D_2 also has the characteristic element (volume, 500 mL). According to Divergence rule 3.1.5, more matter-elements can be expanded to:

$$\big(\text{Teacup } D_1, \quad \text{volume}, \quad 500 \text{ mL}\big)$$
$$\xrightarrow{\text{rule } 3.1.5} \big\vert \big\{ \big(\text{Bowl } D_2, \quad \text{volume}, \quad 500 \text{ mL}\big), \quad \big(\text{Plate } D_3, \quad \text{volume}, \quad 500 \text{ mL}\big),$$
$$\big(\text{Barrel } D_4, \quad \text{volume}, \quad 500 \text{ mL}\big), \quad \big(\text{Basin } D_5, \quad \text{volume}, \quad 500 \text{ mL}\big) \big\}$$

Therefore, when we need to "fill the water" and cannot find a teacup, we may use a bowl to "fill the water." We can choose other items with "volume" as well. This is the reason that "items with the same function can be substituted by each other."

Example 3.1.4

One material in both fireproof board and fireproof paper is asbestos. Therefore both of them have the function of fire prevention. The price of fireproof paper is lower than that of fireproof board. When fireproof board is not available, fireproof paper can be used as a substitute. This is the result of applying Divergence rules 3.1.5, 3.1.1, and 3.1.2, simultaneously. That is,

$$\big(\text{Fireproof board } D_1, \quad \text{material}, \quad \text{asbestos}\big) \xrightarrow{\text{rule } 3.1.5} \big\vert \big(\text{Fireproof paper } D_2, \quad \text{material}, \quad \text{asbestos}\big)$$

$$\big(\text{Fireproof board } D_1, \quad \text{material}, \quad \text{asbestos}\big) \xrightarrow{\text{rule } 3.1.1} \big\vert \big(\text{Fireproof board } D_1, \quad \text{cost}, \quad v_1\big)$$

$$\big(\text{Fireproof board } D_1, \quad \text{cost}, \quad v_1\big) \xrightarrow{\text{rule } 3.1.2} \big\vert \big(\text{Fireproof paper } D_2, \quad \text{cost}, \quad v_2\big)$$

Example 3.1.5

For a product D, the associated actions may be "produce," "transport," "store," "sell," "purchase," etc. According to Divergence rule 3.1.5, starting from "produce product D," we can obtain a divergent tree of the affair-element as follows:

$$A = (\text{Produce, dominating object, product } D)$$

$$\underline{\text{rule 3.1.5}} \left\{ \begin{array}{l} A_1 = (\text{Transport, dominating object, product } D) \\ A_2 = (\text{Store, dominating object, product } D) \\ A_3 = (\text{Sell, dominating object, product } D) \\ A_4 = (\text{Purchase, dominating object, product } D) \end{array} \right.$$

The action "produce" has more characteristics beyond "dominating object" and those corresponding values. According to Divergence rule 3.1.1, we can obtain a divergent tree of the affair-element as follows:

$$A = (\text{Produce, dominating object, product } D)$$

$$\underline{\text{rule 3.1.1}} \left\{ \begin{array}{l} A_1 = (\text{Produce, acting object, enterprise } E) \\ A_2 = (\text{Produce, receiving object, consumer } F) \\ A_3 = (\text{Produce, mode, batch}) \\ A_4 = (\text{Produce, place, Zhongshan}) \\ A_5 = (\text{Produce, number, ten thousand/year}) \end{array} \right.$$

Example 3.1.6

Apply the divergent tree method to analyze "shoes" and people's demand of "wearing shoes" to generate the ideas of broadening the shoe market.

Any pair of shoes can be formalized by a multidimensional matter-element as:

$$M = \begin{bmatrix} \text{Shoe } O_m, & \text{material } c_1, & v_1 \\ & \text{size } c_2, & v_2 \\ & \text{color } c_3, & v_3 \\ & \text{style } c_4, & v_4 \\ & \text{brand } c_5, & v_5 \\ & \text{cost } c_6 & v_6 \\ & \vdots & \vdots \end{bmatrix}$$

According to the divergent tree method, the value of Shoes O_m with regard to each of its characteristic can be extended. Consumers can also buy shoes of different values according to their own demands. So, we have

$$
M_1 = \begin{bmatrix}
\text{Shoe } O_1, & \text{material } c_1, & \text{cattlehide} \\
 & \text{size } c_2, & 40 \\
 & \text{color } c_3, & \text{black} \\
 & \text{style } c_4, & \text{boss style} \\
 & \text{brand } c_5, & \text{FUGUINIAO} \\
 & \text{cost } c_6 & 200 \text{ yuan} \\
 & \vdots & \vdots
\end{bmatrix},
$$

$$
M_2 = \begin{bmatrix}
\text{Shoe } O_2, & \text{material } c_1, & \text{sheepskin} \\
 & \text{size } c_2, & 36 \\
 & \text{color } c_3, & \text{white} \\
 & \text{style } c_4, & \text{leisure style} \\
 & \text{brand } c_5, & \text{FUGUINIAO} \\
 & \text{cost } c_6 & 150 \text{ yuan} \\
 & \vdots & \vdots
\end{bmatrix}, \dots
$$

Enterprises can develop various products according to different demands of different consumers.

For merchants selling shoes, the emphasis is not the divergent analysis of the shoes, but the divergence analysis of consumers' demands, namely, the extensible analysis of consumer's demands of "protecting feet."

The consumer's basic demand for "protecting feet" can be expressed as an affair-element:

$$
A = \begin{bmatrix}
\text{Protect,} & \text{dominating object } c_{a1}, & \text{feet} \\
 & \text{acting object } c_{a2}, & \text{people} \\
 & \text{place } c_{a3}, & \text{on the road} \\
 & \text{time } c_{a4}, & \text{daytime}
\end{bmatrix}
$$

Obviously, there are a lot of shoes satisfying this basic demand. But people's demand for shoes is more than this basic demand, and different people have different demands.

According to the divergence tree method, we can get the following affair-element divergence tree:

$$
A \dashv \left\{
\begin{array}{l}
A_1 = \begin{bmatrix} \text{Protect,} & c_{a1}, & \text{feet} \\ & c_{a2}, & \text{student} \\ & c_{a3}, & \text{playground} \\ & c_{a4}, & \text{daytime} \end{bmatrix} \dashv \left\{
\begin{array}{l}
A_{11} = \begin{bmatrix} \text{Protect,} & c_{a1}, & \text{feet} \\ & c_{a2}, & \text{schoolgirl} \\ & c_{a3}, & \text{on the road} \\ & c_{a4}, & \text{daytime} \end{bmatrix} \\[2em]
A_{12} = \begin{bmatrix} \text{Protect,} & c_{a1}, & \text{feet} \\ & c_{a2}, & \text{schoolboy} \\ & c_{a3}, & \text{playground} \\ & c_{a4}, & \text{daytime} \end{bmatrix}
\end{array} \right. \\[4em]

A_2 = \begin{bmatrix} \text{Show,} & c_{a1}, & \text{position} \\ & c_{a2}, & \text{white-collar class} \\ & c_{a3}, & \text{office} \\ & c_{a4}, & \text{daytime} \end{bmatrix} \dashv A_{21} = \begin{bmatrix} \text{Show,} & c_{a1}, & \text{position} \\ & c_{a2}, & \text{schoolgirl} \\ & c_{a3}, & \text{school} \\ & c_{a4}, & \text{daytime} \end{bmatrix} \dashv A_{211} = \begin{bmatrix} \text{Show,} & c_{a1}, & \text{temperament} \\ & c_{a2}, & \text{schoolgirl} \\ & c_{a3}, & \text{school} \\ & c_{a4}, & \text{daytime} \end{bmatrix} \\[4em]

A_3 = \begin{bmatrix} \text{Defense,} & c_{a1}, & \text{cold} \\ & c_{a2}, & \text{aged people} \\ & c_{a3}, & \text{on the road} \\ & c_{a4}, & \text{winter} \end{bmatrix} \dashv A_{31} = \begin{bmatrix} \text{Defense,} & c_{a1}, & \text{cold} \\ & c_{a2}, & \text{schoolgirl} \\ & c_{a3}, & \text{on the road} \\ & c_{a4}, & \text{winter} \end{bmatrix} \\[4em]

A_4 = \begin{bmatrix} \text{Increase,} & c_{a1}, & \text{height} \\ & c_{a2}, & \text{schoolgirl} \\ & c_{a3}, & \text{school} \\ & c_{a4}, & \text{daytime} \end{bmatrix} \\[4em]

A_5 = \begin{bmatrix} \text{Play,} & c_{a1}, & \text{program} \\ & c_{a2}, & \text{actor} \\ & c_{a3}, & \text{stage} \\ & c_{a4}, & \text{performance time} \end{bmatrix}
\end{array} \right.
$$

Namely

$$
A \dashv \begin{cases} A_1 \dashv \begin{cases} A_{11} \\ A_{12} \end{cases} \\ A_2 \dashv A_{21} \dashv A_{211} \\ A_3 \dashv A_{31} \\ A_4 \\ A_5 \end{cases}
$$

According to this demand divergent tree and data from market investigation, after an investigation and evaluation of each kind of demand, it has been found that the shoes exclusively for the middle school female students, which wear comfortably, are convenient for sports, and can show the temperament of female students, are very promising. Some people have used this analysis result. As a result of accurately finding the market blind spots and avoiding the fierce market competition, they were successful. This is also the so-called "Blue Ocean" approach.

Example 3.1.7

In a lighting product innovation, the parts of any lamp can have a variety of relationships among themselves. Lamp holder D_1 and lampshade D_2, for example, usually have an upper and lower relationship, which can be represented as:

$$
R = \begin{bmatrix} \text{Up and down relation,} & \text{antecedent,} & \text{lampshade } D_2 \\ & \text{consequent,} & \text{lamp holder } D_1 \\ & \text{maintain way,} & \text{spiral} \\ & \text{degree,} & \text{close} \end{bmatrix}
$$

Starting from this relationship element, according to divergence rules, we can obtain the following divergent tree:

$$
R \dashv \begin{cases} R_1 = \begin{bmatrix} \text{Up and down relation,} & \text{antecedent,} & \text{lamp holder } D_1 \\ & \text{consequent,} & \text{lampshade } D_2 \\ & \text{maintain way,} & \text{spiral} \\ & \text{degree,} & \text{close} \end{bmatrix} \\ R_2 = \begin{bmatrix} \text{Left and right relation,} & \text{antecedent,} & \text{lamp holder } D_1 \\ & \text{consequent,} & \text{lampshade } D_2 \\ & \text{maintain way,} & \text{spiral} \\ & \text{degree,} & \text{close} \end{bmatrix} \dashv \\ R_2 = \begin{bmatrix} \text{Up and down relation,} & \text{antecedent,} & \text{lampshade } D_2 \\ & \text{consequent,} & \text{lamp holder } D_1 \\ & \text{maintain way,} & \text{embed} \\ & \text{degree,} & \text{close} \end{bmatrix} \end{cases}
$$

$$R_{21} = \begin{bmatrix} \text{Left and right relation,} & \text{antecedent,} & \text{lamp holder } D_1 \\ & \text{consequent,} & \text{lampshade } D_2 \\ & \text{maintain way,} & \text{embed} \\ & \text{degree,} & \text{loose} \end{bmatrix}$$

$$R_{22} = \begin{bmatrix} \text{Left and right relation,} & \text{antecedent,} & \text{lampshade } D_2 \\ & \text{consequent,} & \text{lamp holder } D_1 \\ & \text{maintain way,} & \text{spiral} \\ & \text{degree,} & \text{close} \end{bmatrix}$$

Thus we can get a lot of innovative ideas of products.

3.2 Correlation Network Method

QUESTIONS AND THINKING:

- How do we identify, analyze, or express relationships and interactions between things?
- In the previous decorative lighting design, the change of which of the object's values about what characteristic can cause the value of this object or the value of another object (and about what characteristic) to change?
- Do the "material" value and "cost" value of a lamp affect each other?
- Do the "voltage" value of the bulb and the "voltage" value of the power supply affect each other?
- Do the light bulb's "brightness" value and the power supply's "voltage" value affect each other?

Pre-knowledge: Correlation Rule

In the objective world, each affair or matter has complicated relationships with other affairs or matters. It is because of the existence of these connections that any transformation of an object causes changes to the objects associated with the object.

Correlation analysis is a method to analyze special relationships between one basic-element and another based on the correlation of matters and affairs. There are three kinds of commonly used correlation rules:

Correlation rule 3.2.1 For two basic-elements with the same object but different characteristics,

$$B_1 = \begin{pmatrix} O, & c_1, & v_1 \end{pmatrix} \quad \text{and} \quad B_2 = \begin{pmatrix} O, & c_2, & v_2 \end{pmatrix}.$$

If there is some kind of functional relationship between their values, namely $v_1 = f_1(v_2)$ or $v_2 = f_2(v_1)$, then B_1 and B_2 are said to be same-object-different-characteristic correlative.

If there is only $v_2 = f_2(v_1)$, the situation is denoted as $B_1 \overset{\sim}{\rightarrow} B_2$. If there is only $v_1 = f_1(v_2)$, the situation is denoted as $B_1 \overset{\sim}{\leftarrow} B_2$. If there are both $v_1 = f_1(v_2)$ and $v_2 = f_2(v_1)$, the situation is denoted as $B_1 \sim B_2$.

Correlation rule 3.2.2 For two basic-elements with different objects and the same characteristic,

$$B_1 = \left(O_1, \quad c, \quad v_1\right) \quad \text{and} \quad B_2 = \left(O_2, \quad c, \quad v_2\right).$$

If there is some kind of functional relationship between their values, namely $v_1 = f_1(v_2)$ or $v_2 = f_2(v_1)$, then B_1 and B_2 are said to be different-object-same-characteristic correlative.

Correlation rule 3.2.3 For two basic-elements with different objects and different characteristics,

$$B_1 = \left(O_1, \quad c_1, \quad v_1\right) \quad \text{and} \quad B_2 = \left(O_2, \quad c_2, \quad v_2\right).$$

If there is some kind of functional relationship between their values, namely $v_1 = f_1(v_2)$ or $v_2 = f_2(v_1)$, then B_1 and B_2 are said to be different-object-different-characteristic correlative.

Most of these correlation rules come from common sense or domain. They can also be obtained from available databases or the knowledge base acquired through data mining.

There are similar correlation rules between complex-elements formed by basic-elements about some evaluation characteristics, which is not detailed in this book.

Instructions: The correlations of basic-elements can be either unidirectional or bidirectional. To avoid causing confusion, unidirectional correlation is usually denoted as $B_1 \overset{\sim}{\rightarrow} B_2$, and bidirectional correlation is usually denoted as $B_1 \sim B_2$. In applications, if there is no special description, the symbol "~" is used to denote the correlation. The symbol "$\overset{\sim}{\rightarrow}$" is applied only when the direction needs to be specified.

A correlation can also be either an "AND correlation" or an "OR correlation." According to different situations, we can use the following symbols to denote them:

① "AND correlation" between one basic-element B and multiple basic-elements $B_1, \ldots, B_m$ is shown as $B \sim \overset{m}{\underset{i=1}{\wedge}} B_i$

② "OR correlation" between one basic-element B and multiple basic-elements $B_1, \ldots, B_m$" is shown as $B \sim \overset{m}{\underset{i=1}{\vee}} B_i$

③ "Unidirectional AND correlation" between one basic-element B and multiple basic-elements $B_1, \ldots, B_m$ is shown as $B \overset{\sim}{\rightarrow} \overset{m}{\underset{i=1}{\wedge}} B_i$

④ "Unidirectional OR correlation" between one basic-element B and multiple basic-elements $B_1, \ldots, B_m$ is shown as $B \overset{\sim}{\to} \overset{m}{\underset{i=1}{\vee}} B_i$

⑤ "Unidirectional AND correlation" between multiple basic-elements $B_1, \ldots, B_m$ and one basic-element B is shown as $\overset{m}{\underset{i=1}{\wedge}} B_i \overset{\sim}{\to} B$

⑥ "Unidirectional OR correlation" between multiple basic-elements $B_1, \ldots, B_m$ and one basic-element B'' is shown as $\overset{m}{\underset{i=1}{\vee}} B_i \overset{\sim}{\to} B$

Correlation situations between multiple basic-elements and multiple basic-elements also have similar results, which are not detailed here.

Basic Steps of the Correlation Network Method

According to these correlation rules, formalized methods can be used to describe such a correlative relationship between basic-elements. Because the relationships between a basic-element and other basic-elements are like a network structure, it is called a correlation network. A correlation tree is a special instance of a correlation networks. We use symbols to illustrate the concept of correlation networks as in Figure 3.2.1.

In a correlation network, a change of one basic-element can result in changes of other associated basic-elements involved in the network. Generally speaking, each correlation network is dynamic. But at a given moment, for a given basic-element, its correlation network is uniquely determined.

The method of finding solutions to contradictory problems through a correlation network is called the correlation network method. The basic steps are as follows:

1. Write down the basic-element B for analysis;
2. Use correlation rules to construct the correlation network of the basic-element B;

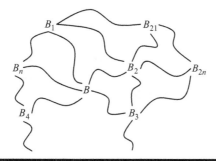

Figure 3.2.1 Schematic diagram of a correlation network.

3. Analyze the correlation network to determine the basic-element B_j that causes the basic-element B to change, or basic-elements B_j that are changed due to changes of the basic-element B;
4. Optionally apply the basic-element B_j in the correlation network to innovate or solve the given contradictory problem.

Explanation: When innovating or solving a contradictory problem, sometimes we can compulsively relieve correlative relationships or establish correlative relationships. This is also an important means to innovate or solve the contradictory problems. If one can cautiously apply it, she will get good results.

Case Analysis:

Example 3.2.1

According to the divergence rules, a light bulb has many characteristics and corresponding values. Among the characteristics, the value of one characteristic may be related to that of another characteristic. These relationships form the same-object-different-characteristic correlation tree. According to the domain knowledge, there is a functional relationship between the "brightness" and "wattage" values of the bulb. According to Correlation rule 3.2.1, we have

$$(\text{Bulb } D_1, \quad \text{brightness}, \quad v_{11}) \sim (\text{Bulb } D_1, \quad \text{wattage}, \quad v_{21}).$$

For a chandelier, according to domain knowledge, its "material" value is related to its "weight" value and "cost" value. So, there is the following AND correlation tree:

$$(\text{Chandelier } D_2, \quad \text{material}, \quad v_{12}) \sim (\text{Chandelier } D_2, \quad \text{weight}, \quad v_{22}) \wedge (\text{Chandelier } D_2, \quad \text{cost}, \quad v_{32}).$$

Example 3.2.2

According to the domain knowledge, a chosen bulb and the electrical source have a functional relation about the "voltage" value; a ceiling lamp and the ceiling have a functional relation about the "height" value of distance from the ground; the "length" value of the light tube frame has a functional relation with the "length" value of the light tube. According to Correlation rules 3.2.2, we have

$$(\text{Bulb } D_1, \quad \text{voltage}, \quad v_{13}) \sim (\text{Power supply } D_2, \quad \text{voltage}, \quad v_{23}),$$
$$(\text{Ceiling lamp } D_3, \quad \text{height}, \quad v_{14}) \sim (\text{Ceiling } D_4, \quad \text{height}, \quad v_{24}), \quad \text{and}$$
$$(\text{Light tube frame } D_5, \quad \text{length}, \quad v_{15}) \sim (\text{Light tube } D_6, \quad \text{length}, \quad v_{25}).$$

Example 3.2.3

According to the domain knowledge, there is a functional relationship between the "brightness" value of a light bulb and the "voltage" value of the electrical source.

The "size" value of a chandelier has a functional relationship with the "area" value of a room; the "light intensity" value of a lamp has a functional relationship with the "material" value of the lampshade. According to Correlation rule 3.2.3, we have

$$(\text{Bulb } D_1, \quad \text{brightness}, \quad v_1) \sim (\text{Power supply } D_2, \quad \text{voltage}, \quad v_2),$$
$$(\text{Chandelier } D_3, \quad \text{size}, \quad v_3) \sim (\text{Room } D_4, \quad \text{area}, \quad v_4), \quad \text{and}$$
$$(\text{Lamp } D_5, \quad \text{light intensity}, \quad v_5) \sim (\text{Lampshade } D_6, \quad \text{material}, \quad v_6).$$

Example 3.2.4

Use the correlation network method to analyze the influence of the size of immigratory population to City O_1 on different aspects of the city.

Set $M_1 = (\text{City } O_1, \quad \text{immigratory population number } c_1, \quad v_1) = (O_1, \quad c_1, \quad v_1)$. According to professional knowledge and statistics knowledge, M_1 has the following correlationships:

$$M_1 \sim \begin{cases} M_2 = (O_1, \quad \text{total population } c_2, \quad v_2) \\ M_3 = (O_1, \quad \text{employment opportunity } c_3, \quad v_3) \\ M_4 = (O_1, \quad \text{economic growth rate } c_4, \quad v_4) \\ M_5 = (O_1, \quad \text{quantity of employment } c_5, \quad v_5) \end{cases}$$

and

$$M_3 \sim M_5 \begin{cases} M_{51} = (\text{Construction industry } O_{51}, c_5, v_{51}) \\ \quad \sim M_{511} = (\text{Construction industry } O_{51}, \text{urban construction work } c_{51}, v_{511}) \\ M_{52} = (\text{Service industry } O_{52}, c_5, v_{52}) \\ \quad \sim M_{521} = (\text{Service industry } O_{52}, \text{service project number } c_{52}, v_{521}) \\ M_{53} = (\text{Commerce } O_{53}, c_5, v_{53}) \sim M_{531} = (\text{Commerce } O_{53}, \text{activeness } c_{53}, v_{531}) \end{cases} \sim M_4$$

$$M_2 \sim \begin{cases} M_{21} = (O_1, \quad \text{housing demand } c_{21}, \quad v_{21}) \sim M_{511} \\ M_{22} = (O_1, \quad \text{the demand for school places } c_{22}, \quad v_{22}) \\ \quad \sim M_{221} = (O_1, \quad \text{demand for teachers } c_{221}, \quad v_{221}) \\ M_{23} = (O_1, \quad \text{transportation demand } c_{23}, \quad v_{23}) \sim M_{231} = (O_1, \quad \text{road congestion level } c_{231}, v_{231}) \\ M_{24} = (O_1, \quad \text{diet demand } c_{24}, \quad v_{24}) \sim M_{521} \end{cases}$$

From this analysis, we obtain the following correlation network:

$$M_1 \sim \begin{cases} M_2 \sim \begin{cases} M_{21} \sim M_{511} \\ M_{22} \sim M_{221} \\ M_{23} \sim M_{231} \\ M_{24} \sim M_{521} \end{cases} \\ \\ M_3 \sim M_5 \sim \begin{cases} M_{51} \sim M_{511} \\ M_{52} \sim M_{521} \\ M_{53} \sim M_{531} \end{cases} \sim M_4 \end{cases} .$$

When solving various contradictory problems, city managers must pay attention to all kinds of correlation networks. Otherwise, after solving a contradictory problem, another contradictory problem may emerge.

3.3 Implication System Method

QUESTIONS AND THINKING:

"Relieve the state of Zhao by besieging the state of Wei"—During the Warring States Period, State Qi forced State Wei to withdraw its attack on State Zhao's troops by besieging the capital city of Wei, making Zhao safe. Later, the phrase is used to refer to attacking the enemy's rear stronghold to force the enemy to retreat.

■ Why can one achieve the goal of "relieving Zhao" through "besieging Wei"?
■ Can we think of situations that can be dealt with in this way?
■ When a goal is not easy to achieve, what do people usually do?

Pre-knowledge: Implication and Implication Rule

Implication analysis is a formalized method developed for studying basic-elements based on implication properties among matters, affairs, and relations.

Implication includes causal implication and existence implication. It can also be categorized into unconditional implication and conditional implication.

Causal implication: Suppose B_1, B_2 are two basic-elements. If the implementation of B_1 always leads to the implementation of B_2, then basic-element B_1 is said to imply basic-element B_2, denoted as $B_1 \Rightarrow B_2$. If B_1's implementation leads to B_2's implementation under condition l, we say, "Under condition l, B_1 implies B_2, denoted as $B_1 \Rightarrow (l) B_2$."

No matter whether $B_1 \Rightarrow B_2$, or $B_1 \Rightarrow (l) B_2$, we usually call B_1 the inferior basic-element and B_2 as the superior basic-element.

Existence implication: Suppose B_1, B_2 are two basic-elements. If the existence of B_1 always leads to the existence of B_2, then basic-element B_1 is said to imply basic-element B_2, denoted as $B_1 \Rightarrow B_2$. If B_1's existence leads to B_2's existence under condition l, then we say, "Under condition l, basic-element B_1 implies basic-element B_2, denoted as $B_1 \Rightarrow (l) B_1$."

Each existence implication is mainly a matter-element implication and a relation-element implication. Each causal implication is mainly an affair-element implication. This includes implications of goal affair-element, functional affair-element, demand affair-element, transformation, and so on.

Implication rule 3.3.1 Suppose that we have basic-elements B, B_1, and B_2.

  a. If B_1's implementation and B_2's implementation lead to B's implementation, then B_1, B_2 AND imply B, denoted as $B_1 \wedge B_2 \Rightarrow B$.
  b. If B_1's implementation or B_2's implementation leads to B's implementation, then B_1, B_2 OR imply B, denoted as $B_1 \vee B_2 \Rightarrow B$.
  c. If B's implementation leads to both B_1's implementation and B_2's implementation, then B AND implies B_1, B_2, denoted as $B \Rightarrow B_1 \wedge B_2$.
  d. If B's implementation leads to B_1's implementation or B_2's implementation, then B OR implies B_1, B_2, denoted as $B \Rightarrow B_1 \vee B_2$. This is similar to (c).

These rules can also be extended to more general situations:

$$B \Rightarrow \overset{n}{\underset{i=1}{\wedge}} B_i, \quad B \Rightarrow \overset{n}{\underset{i=1}{\vee}} B_i, \quad \overset{n}{\underset{i=1}{\wedge}} B_i \Rightarrow B, \quad \overset{n}{\underset{i=1}{\vee}} B_i \Rightarrow B.$$

Implication rule 3.3.2 If $B_1 \Rightarrow B_2$, $B_2 \Rightarrow B_3$, then $B_1 \Rightarrow B_3$, which can also be denoted as $B_1 \Rightarrow B_2 \Rightarrow B_3$.

Implication rule 3.3.3 If $B_{11} \wedge B_{12} \Rightarrow B_1$, $B_{21} \wedge B_{22} \Rightarrow B_2$, and $B_1 \wedge B_2 \Rightarrow B$, then $B_{11} \wedge B_{12} \wedge B_{21} \wedge B_{22} \Rightarrow B$.

This rule shows that, in the AND implication, the most inferior basic-elements jointly imply the top-level basic-element.

Implication rule 3.3.4 If $B_{11} \vee B_{12} \Rightarrow B_1$, $B_{21} \vee B_{22} \Rightarrow B_2$, and $B_1 \vee B_2 \Rightarrow B$, then $B_{11} \vee B_{12} \vee B_{21} \vee B_{22} \Rightarrow B$.

This rule shows that, in the OR implication, each lowest-level basic-element implies the top-level basic-element.

The system formed by these rules is called the basic-element implication system. The general form of basic-element implication system is (Figure 3.3.1):

This implication system can be an "AND implication system," an "OR implication system," or an "AND/OR implication system." This shows that the implication system can be multi-layered. When the upper basic-element is not easy to implement, we can look for its inferior basic-element. If the lower basic-element is easy to implement, it is considered that a path to innovation or solving contradictory problems is found.

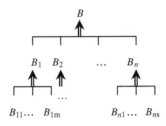

Figure 3.3.1 The general form of basic-element implication system.

Basic Steps of the Implication System Method

The implication system method is used to analyze a basic-element based on the implication rules to find ways to innovate or solve contradictory problems. The basic steps of using this method are as follows:

1. List the basic-element, transformation, or problem to be analyzed;
2. Based on the domain knowledge, common sense knowledge and implication rules, establish the implication system;
3. According to the information newly emerging in the problem-solving process, add or truncate the implication system on some layer of the system. If no new information appears, then go to the next step;
4. By implementing the most inferior basic-elements first, implement the most superior basic-element, and find a path to innovation or solution of the given contradictory problem.

No matter what kind of implication system is given, it can be an "AND implication system", "OR implication system" or "AND/OR implication system". In a specific application, we must notice the differences among these systems.

Using the implication system method, when some basic-element is not easy to implement, we may then seek its inferior basic-elements. As long as one of them can be implemented, we can achieve the ultimate goal and resolve the situation of concern. Similar to the correlative network approach, compulsively establishing or eliminating implication relationships can be carried out when innovating or resolving contradictory problems.

Case Analysis:

Example 3.3.1

For a chandelier with a weight of 1,000 g, its lamp holder, lamp bracket, etc., are all parts. If the weight of the lamp holder is 300 g and the weight of the lamp bracket is 500 g, there is the following existence implication system of matter-elements:

$$\left(\text{Chandelier } D, \quad \text{weight}, \quad 1000 \text{ g}\right)$$

$$\Rightarrow \left(\text{Lamp holder } D_1, \quad \text{weight}, \quad 300 \text{ g}\right) \wedge \left(\text{Lamp bracket } D_2, \quad \text{weight}, \quad 500 \text{ g}\right)$$

Example 3.3.2

For a wall lamp, if it gives out heat, it can heat the wall. This is the causal implication of the wall lamp's function, and we can develop the following implication system of the affair-elements:

$$
\begin{bmatrix}
\text{Provide,} & \text{dominating object,} & \text{heat} \\
& \text{tool,} & \text{wall lamp } D
\end{bmatrix}
$$

$$
\Rightarrow
\begin{bmatrix}
\text{Heat,} & \text{dominating object,} & \text{wall} \\
& \text{tool,} & \text{wall lamp } D
\end{bmatrix}
$$

Example 3.3.3

If an enterprise's goal is to "increase sales," then according to the relevant domain knowledge, the following implication analysis can be performed on realizing this goal (Figure 3.3.2):

This is an implication system of the goal, which can be further detailed by the implication system of affair-elements as follows:

$$
\begin{bmatrix}
\begin{bmatrix}
\text{Improve,} & \text{dominating object,} & \text{quantity of sale} \\
& \text{receiving object,} & \text{product } D \\
& \text{degree,} & 10\% \\
& \text{time,} & 1 \text{ year}
\end{bmatrix} \\
\Leftarrow
\begin{cases}
\begin{bmatrix}
\begin{bmatrix}
\text{Improve,} & \text{dominating object,} & \text{quality} \\
& \text{receiving object,} & \text{product } D
\end{bmatrix} \\
\overset{\wedge}{\Leftarrow}
\begin{cases}
\begin{bmatrix}
\text{Improve,} & \text{dominating object,} & \text{design level} \\
& \text{receiving object,} & \text{product } D
\end{bmatrix} \\
\begin{bmatrix}
\text{Improve,} & \text{dominating object,} & \text{technological level} \\
& \text{receiving object,} & \text{product } D
\end{bmatrix}
\end{cases}
\end{bmatrix} \\
\begin{bmatrix}
\begin{bmatrix}
\text{Improve,} & \text{dominating object,} & \text{service level} \\
& \text{degree,} & 50\%
\end{bmatrix} \\
\Leftarrow
\begin{bmatrix}
\text{Increase,} & \text{dominating object,} & \text{warranty period} \\
& \text{time,} & 5 \text{ years}
\end{bmatrix}
\end{bmatrix} \\
\begin{bmatrix}
\begin{bmatrix}
\text{Reduce,} & \text{dominating object,} & \text{price} \\
& \text{degree,} & 5\%
\end{bmatrix} \\
\overset{\wedge}{\Leftarrow}
\begin{cases}
\begin{bmatrix}
\text{Reduce,} & \text{dominating object,} & \text{cost} \\
& \text{degree,} & 10\%
\end{bmatrix} \\
\begin{bmatrix}
\text{Decrease,} & \text{dominating object,} & \text{process} \\
& \text{degree,} & 20\%
\end{bmatrix}
\end{cases}
\end{bmatrix} \\
\begin{bmatrix}
\text{Increase,} & \text{dominating object,} & \text{varieties} \\
& \text{number,} & 5
\end{bmatrix}
\end{cases}
\end{bmatrix}
$$

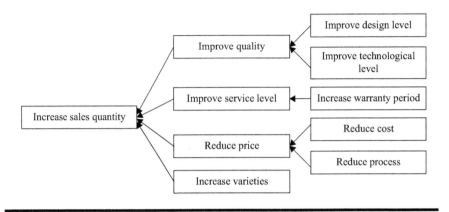

Figure 3.3.2   Implication analysis of "increase sales quantity".

Explanation: The symbols $\overset{\wedge}{\Leftarrow}$ and $\overset{\wedge}{\Rightarrow}$ mean the AND implication. The symbols $\overset{\vee}{\Leftarrow}$ and $\overset{\vee}{\Rightarrow}$ mean the OR implication. In a specific application, we should use them according to the actual situation.

EXTENDED KNOWLEDGE

The relationship and difference between function and effect

1. Function and effect are two concepts that are interrelated while different from each other.
2. The concept of function is the intrinsic efficiency of things, determined by the internal factoral structure of things. It is a relatively stable and independent mechanism within the things.
3. The concept of effect is the external effect of the relationship between the thing of concern and the external environment.
4. The effect of a function on the outside world may be positive or negative, depending on how the function and the external environment interact.
5. Generally speaking, function is the internal basis and premise of effect; objective need is the external condition of the effect evaluation; the effect is the actual efficiency generated by a combination of the function and the objective demand.

Example 3.3.4

Consider the function of a light bulb. It can generate heat and has the following function implication system, of which some effects are positive effects, while the other effects are negative (Figure 3.3.3).

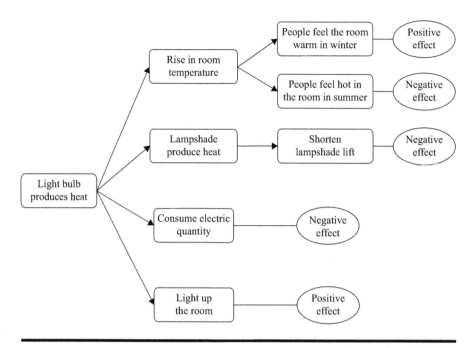

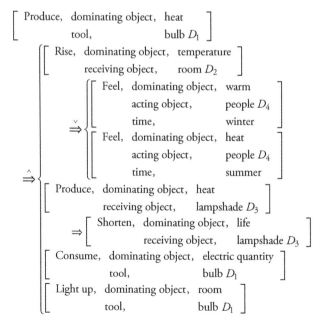

Figure 3.3.3   Implication analysis of "light bulb produces heat".

The implication system of this function can also be further detailed by the following implication system of affair-elements:

$$
\begin{bmatrix}
\text{Produce,} & \text{dominating object,} & \text{heat} \\
& \text{tool,} & \text{bulb } D_1
\end{bmatrix}
$$

$$
\stackrel{\wedge}{\Rightarrow}
\begin{cases}
\begin{bmatrix}
\text{Rise,} & \text{dominating object,} & \text{temperature} \\
& \text{receiving object,} & \text{room } D_2
\end{bmatrix} \\
\qquad \stackrel{\vee}{\Rightarrow}
\begin{cases}
\begin{bmatrix}
\text{Feel,} & \text{dominating object,} & \text{warm} \\
& \text{acting object,} & \text{people } D_4 \\
& \text{time,} & \text{winter}
\end{bmatrix} \\
\begin{bmatrix}
\text{Feel,} & \text{dominating object,} & \text{heat} \\
& \text{acting object,} & \text{people } D_4 \\
& \text{time,} & \text{summer}
\end{bmatrix}
\end{cases} \\
\begin{bmatrix}
\text{Produce,} & \text{dominating object,} & \text{heat} \\
& \text{receiving object,} & \text{lampshade } D_3
\end{bmatrix} \\
\qquad \Rightarrow
\begin{bmatrix}
\text{Shorten,} & \text{dominating object,} & \text{life} \\
& \text{receiving object,} & \text{lampshade } D_3
\end{bmatrix} \\
\begin{bmatrix}
\text{Consume,} & \text{dominating object,} & \text{electric quantity} \\
& \text{tool,} & \text{bulb } D_1
\end{bmatrix} \\
\begin{bmatrix}
\text{Light up,} & \text{dominating object,} & \text{room} \\
& \text{tool,} & \text{bulb } D_1
\end{bmatrix}
\end{cases}
$$

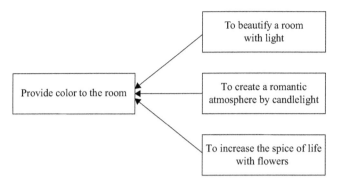

Figure 3.3.4   Implication analysis of "provide color to the room".

Example 3.3.5

If a customer needs to "provide color for a room," we can perform an implication on the customer's demands by obtaining the following implication system (Figure 3.3.4):

The implication system of demands can also be further detailed by the following implication system of affair-elements:

$$
\begin{bmatrix}
\text{Provide,} & \text{dominating object,} & \text{color} \\
& \text{receiving object,} & \text{room } D_1
\end{bmatrix}
$$

$$
\overset{\vee}{\Leftarrow}
\begin{cases}
\begin{bmatrix}
\text{Beautify,} & \text{dominating object,} & \text{room } D_1 \\
& \text{mode,} & \text{light}
\end{bmatrix} \\[2ex]
\begin{bmatrix}
\text{Make,} & \text{dominating object,} & \text{romantic atmosphere} \\
& \text{mode,} & \text{candlelight}
\end{bmatrix} \\[2ex]
\begin{bmatrix}
\text{Increase,} & \text{dominating object,} & \text{spice of life} \\
& \text{mode,} & \text{flower}
\end{bmatrix}
\end{cases}
$$

Example 3.3.6

The light source type of a lamp has a series of related relationships with the lamp structure, production process, and costs. According to the domain knowledge and the correlation network method of Section 3.2, we have the following correlation network of matter-elements:

$$(\text{Lamp } D, \text{light type}, v_1) \sim (\text{Lamp } D, \text{structure}, v_2)$$
$$\sim (\text{Lamp } D, \text{manufacturing process}, v_3) \sim (\text{Lamp } D, \text{cost}, v_4)$$

When we want to change the light source type of the lamp, the other matter-elements in the correlation network will also change accordingly. The result of the change forms the implication system of transformations (essentially the conductive transformation, which will be described in detail in Chapter 5).

Example 3.3.7

If D_1 is D_2's grandfather, D_3 is D_1's son, and D_3 is D_2's father, then there must be the following implication system of the relation-elements:

$$\begin{bmatrix} \text{Grandparents-and-grandchild relation,} & \text{antecedent,} & D_1 \\ & \text{consequent,} & D_2 \end{bmatrix}$$

$$\overset{\wedge}{\Rightarrow} \begin{cases} \begin{bmatrix} \text{Parent child relation,} & \text{antecedent,} & D_1 \\ & \text{consequent,} & D_3 \end{bmatrix} \\ \begin{bmatrix} \text{Parent child relation,} & \text{antecedent,} & D_3 \\ & \text{consequent,} & D_2 \end{bmatrix} \end{cases}$$

For a given table lamp, if the lampshade and the lamp holder have the conventional up-down relationship, then the lamp holder and the light bulb also have the up-down relationship, namely the following implication system of relation-elements:

$$\begin{bmatrix} \text{Up and down relation,} & \text{antecedent,} & \text{lampshade } D_1 \\ & \text{consequent,} & \text{lamp holder } D_2 \end{bmatrix}$$

$$\Rightarrow \begin{bmatrix} \text{Up and down relation,} & \text{antecedent,} & \text{bulb } D_1 \\ & \text{consequent,} & \text{lamp holder } D_2 \end{bmatrix}$$

Example 3.3.8

Under the condition that lamp D_1 is located in room D_2, namely, $l = (\text{lamp } D_1, \text{location, room } D_2)$, we have the following condition implication system of affair-elements:

$$\begin{bmatrix} \text{Open,} & \text{dominating object,} & \text{lamp } D_1 \\ & \text{acting object,} & \text{people } D_3 \end{bmatrix} \underset{(l)}{\Rightarrow} \begin{bmatrix} \text{Light up,} & \text{dominating object,} & \text{room } D_2 \\ & \text{tool,} & \text{lamp } D_1 \end{bmatrix}$$

$$\underset{(l)}{\Rightarrow} \begin{bmatrix} \text{Have,} & \text{dominating object,} & \text{light ray} \\ & \text{location,} & \text{room } D_2 \end{bmatrix} \underset{(l)}{\Rightarrow} \begin{bmatrix} \text{Read,} & \text{dominating object,} & \text{book} \\ & \text{acting object,} & \text{people } D_3 \\ & \text{place,} & \text{room } D_2 \\ & \text{time,} & \text{evening} \end{bmatrix}$$

Example 3.3.9

For the lighting industry, facing the consumers' demand of "decorating rooms," we can perform an implication analysis on the demand to find more specific demands and then create more targeted products to satisfy the consumers' demand. For instance,

$$
\left(\text{Decorate}, \quad \text{dominating object}, \quad \text{room } D\right)
$$

$$
\Leftarrow \vee \left\{
\begin{array}{l}
\left[\begin{array}{ccc}
\text{Decorate}, & \text{dominating object}, & \text{room } D_1 \\
& \text{way}, & \text{color}
\end{array}\right] \\
\quad \Leftarrow \left[\begin{array}{ccc}
\text{Install}, & \text{dominating object}, & \text{colored lantern } D_2 \\
& \text{place}, & \text{room } D_1
\end{array}\right] \\
\left[\begin{array}{ccc}
\text{Decorate}, & \text{dominating object}, & \text{room } D_1 \\
& \text{way}, & \text{fabrics}
\end{array}\right] \\
\quad \Leftarrow \left[\begin{array}{ccc}
\text{Install}, & \text{dominating object}, & \text{lampshade } D_3 \\
& \text{place}, & \text{room } D_1
\end{array}\right] \\
\left[\begin{array}{ccc}
\text{Decorate}, & \text{dominating object}, & \text{room } D_1 \\
& \text{way}, & \text{light ray}
\end{array}\right] \\
\quad \Leftarrow \left[\begin{array}{ccc}
\text{Install}, & \text{dominating object}, & \text{illuminant } D_4 \\
& \text{place}, & \text{room } D_1
\end{array}\right]
\end{array}
\right.
$$

3.4 Decomposition/Combination Chain Method

QUESTIONS AND THINKING:

■ How does a person measure the thickness of a thin piece of paper with a mm-scaled ruler?
■ Why do people who recycle old electrical appliances often disassemble and sell the old ones?
■ At school season, manufacturers of school supplies often place a variety of stationeries bundled and preferentially sell them. What is the truth behind the sale?
■ If the cartoon character in a cartoon image is made as large as a live person and placed in a playground, then what effect does the action have?

Such methods are often used in innovating or solving contradictory problems.

Pre-knowledge: Opening-up Rule

The possibility that affairs, matters, and relations can be combined, decomposed, and enlarged is termed as composability, decomposability, and expandability/contractibility, respectively, and these are known collectively as scalability.

According to composability, a thing can be combined with other things to generate new things, thus providing the possibility of solving contradictory problems. According to decomposability, a thing can be decomposed into some new things, which have some characteristics that the original thing does not have, thus providing the possibility of solving contradictory problems. Similarly, a thing can be expanded or contracted to provide possibilities of solving contradictory problems.

Affairs, matters, and relations are represented by basic-elements, on which the so-called opening-up analysis can be performed, including composability, decomposability, and expandability/contractibility analyses.

Opening-up rule 3.4.1 (Composability rules) Given basic-element $B_1 = (O_1, c_1, v_1)$, there is at least one basic-element $B_2 = (O_2, c_2, v_2)$ so that B_1 and B_2 can be combined into B, where B_2 is called a composable basic-element of B_1, in this case $B = B_1 \oplus B_2$.

$$
=\begin{cases}
(O_1, c_1 \oplus c_2, v_1 \oplus v_2) = \begin{bmatrix} O_1, & c_1, & v_1 \\ & c_2, & v_2 \end{bmatrix}, & O_1 = O_2, \quad c_1 \neq c_2; \\[2ex]
(O_1 \oplus O_2, c_1, v_1 \oplus v_2), & O_1 \neq O_2, \quad c_1 = c_2; \\[2ex]
\begin{bmatrix} O_1 \oplus O_2, & c_1, & v_1 \oplus c_1(O_2) \\ & c_2, & c_2(O_1) \oplus v_2 \end{bmatrix}, & O_1 \neq O_2, \quad c_1 \neq c_2.
\end{cases}
$$

For matter-elements, there are two forms of composability rules:

(1) Increasable rule: Given matter-element $M_1 = (O_{m1}, c_{m1}, c_{m1}(O_{m1}))$, there is at least one matter-element $M_2 = (O_{m2}, c_{m2}, c_{m2}(O_{m2}))$ so that O_{m1} and O_{m2} can form a polymer; namely, O_{m1} and O_{m2} are increasable matters, making

$$
M_1 \oplus M_2 = \begin{cases}
\left(O_{m1} \oplus O_{m2}, \ c_{m1}, \ c_{m1}(O_{m1}) \oplus c_{m1}(O_{m2})\right), & c_{m1} = c_{m2} \\[2ex]
\begin{bmatrix} O_{m1} \oplus O_{m2}, & c_{m1}, & c_{m1}(O_{m1}) \oplus c_{m1}(O_{m2}) \\ & c_{m2}, & c_{m2}(O_{m1}) \oplus c_{m2}(O_{m2}) \end{bmatrix}, & c_{m1} \neq c_{m2}.
\end{cases}
$$

In particular, when $O_{m1} = O_{m2}$ and $c_{m1} \neq c_{m2}$, we have

$$
M_1 \oplus M_2 = \begin{bmatrix} O_{m1}, & c_{m1}, & c_{m1}(O_{m1}) \\ & c_{m2}, & c_{m2}(O_{m1}) \end{bmatrix}.
$$

Namely, two matter-elements with the same matter and different characteristics can form a two-dimensional matter-element.

The matter-element increasable rule illustrates that when a matter-element cannot solve the problem of concern, we should consider combining it with another matter-element. If M_1 cannot attract customers, and M_2 can give customers a surprise, then $M_1 \oplus M_2$ can achieve promotional purposes. An increasable matter-element can be obtained according to the divergent rules.

According to this rule, if a simple application of a matter-element cannot solve the given contradictory problem, one can consider using the aggregation of multiple matter-elements to attack the contradictory problem.

(2) Integrable rule: Given a matter-element $M_1 = (O_{m1}, c_{m1}, c_{m1}(O_{m1}))$, there is at least one same dimensional matter-element $M_2 = (O_{m2}, c_{m2}, c_{m2}(O_{m2}))$ such that M_1 and M_2 constitute a new matter–element, where O_{m1} and O_{m2} can form a system. In this case, we say that O_{m1} and O_{m2} are integrable, namely

$$
M_1 \otimes M_2 = \begin{cases} \left(O_{m1} \otimes O_{m2}, \quad c_{m1} \otimes c_{m2}, \quad c_{m1}(O_{m1}) \otimes c_{m2}(O_{m2}) \right), & \text{when } c_{m1} \text{ and } \\ & \qquad c_{m2} \text{ are integrable} \\[2ex] \left[\begin{array}{ccc} O_{m1} \otimes O_{m2}, & c_{m1}, & c_{m1}(O_{m1} \otimes O_{m2}) \\ & c_{m2}, & c_{m2}(O_{m1} \otimes O_{m2}) \end{array} \right], & \text{when } c_{m1} \text{ and} \\ & \qquad c_{m2} \text{ are not integrable} \end{cases}.
$$

When $O_{m1} \neq O_{m2}$, and $c_{m1} = c_{m2}$, we have

$$
M_1 \otimes M_2 = \left(O_{m1} \otimes O_{m2}, \quad c_{m1}, \quad c_{m1}(O_{m1}) \otimes c_{m1}(O_{m2}) \right)
$$
$$
\underline{\underline{\Delta}} \left(O_{m1} \otimes O_{m2}, \quad c_{m1}, \quad c_{m1}(O_{m1} \otimes O_{m2}) \right)
$$

where $c_{m1}(O_{m1} \otimes O_{m2}) = c_{m1}(O_{m1}) \otimes c_{m1}(O_{m2})$.

When $O_{m1} = O_{m2}$ and $c_{m1} \neq c_{m2}$, we have

$$
M_1 \otimes M_2 = \begin{cases} \left(O_{m1}, \quad c_{m1} \otimes c_{m2}, \quad c_{m1}(O_{m1}) \otimes c_{m2}(O_{m1}) \right), & \text{when } c_{m1} \text{ and} \\ & \qquad c_{m2} \text{ are integrable} \\[2ex] \left[\begin{array}{ccc} O_{m1}, & c_{m1}, & c_{m1}(O_{m1}) \\ & c_{m2}, & c_{m2}(O_{m1}) \end{array} \right], & \text{when } c_{m1} \text{ and} \\ & \qquad c_{m2} \text{ are not integrable} \end{cases}.
$$

Increasable and integrable rules of a matter-element represent two forms of combining patterns. The essence of the increasable rule is aggregation, and the essence of the integrable rule is forming a system.

The integrable rule is also the theoretical basis for our usual practice of "filling the insufficient with surplus." For a disadvantaged condition, through divergence rules we can find advantageous conditions that can be combined with the disadvantaged condition so as to resolve contradictions.

In addition, according to the integrable rule, in a system composed of n matter-elements with the same characteristics, the effect of the combination can be judged by the values of the characteristic before and after the combination. Namely, if

$$M_1 = \left(O_{m1}, \quad c_m, \quad c_m(O_{m1})\right), \quad M_2 = \left(O_{m2}, \quad c_m, \quad c_m(O_{m2})\right), \ldots,$$
$$M_n = \left(O_{mn}, \quad c_m, \quad c_m(O_{mn})\right),$$

then

$$M_1 \otimes M_2 \otimes \cdots \otimes M_n$$
$$= \left(O_{m1} \otimes O_{m2} \otimes \cdots \otimes O_{mn}, \quad c_m, \quad c_m(O_{m1}) \otimes c_m(O_{m2}) \otimes \cdots \otimes c_m(O_{mn})\right)$$
$$\underline{\underline{\Delta}}\left(O_m, \quad c_m, \quad c_m(O_m)\right)$$

The relationship between $c_m(O_m)$ and the sum of the original values $\sum_{i=1}^{n} c_m(O_{mi})$ can have the following three situations:

a. If $c_m(O_m) > \sum_{i=1}^{n} c_m(O_{mi})$, the value after the combination is greater than the sum of the original values;
b. If $c_m(O_m) = \sum_{i=1}^{n} c_m(O_{mi})$, the value after the combination is equal to the sum of the original values; or
c. If $c_m(O_m) < \sum_{i=1}^{n} c_m(O_{mi})$, the value after the combination is less than the sum of the original values.

This also explains that the different results of "two heads are better than one" and "three boys are no boy" are due to the internal relations after the combination. Additional deepened reasons will be interpreted in Chapter 4.

Opening-up rule 3.4.2 (Decomposability rule) Some basic-elements can be decomposed into several basic-elements according to certain conditions. Namely, suppose $B = (O, c, c(O))$, $B_i = (O_i, c, c(O_i))$, $i = 1, 2, \ldots, n$, then under certain condition ℓ, for some characteristic c, we have:

$$(O, c, c(O)) // (\ell)\{(O_1, c, c(O_1)), (O_2, c, c(O_2)), \ldots, (O_m, c, c(O_m))\}$$

denoted as $B // (\ell) \{B_1, B_2, \ldots, B_m\}$.

Obviously, different conditions have different forms of decomposition, namely, B can be decomposed into multiple groups of basic-elements as $B_i = \{B_{i1}, B_{i2}, \ldots, B_{im_i}\}$, $i = 1, 2, \ldots, n$.

Using the decomposability rule, if $O_m = O_{m1} \otimes O_{m2} \otimes \cdots \otimes O_{mn}$, that is, if O_m is a decomposable matter, $O_m // \{O_{m1}, O_{m2}, \ldots O_{mn}\}$, then for any characteristic c_m, we have:

$$\left(O_m, \quad c_m, \quad c_m(O_m)\right) //$$

$$\left\{\left(O_{m1}, \quad c_m, \quad c_m(O_{m1})\right), \left(O_{m2}, \quad c_m, \quad c_m(O_{m2})\right), \ldots, \left(O_{mn}, \quad c_m, \quad c_m(O_{mn})\right)\right\},$$

and the relationship between the value of the matter before the decomposition $c_m(O_m)$ and the sum of the values of the things after the decomposition $\sum_{i=1}^{n} c_m(O_{mi})$ can satisfy one of the following three situations:

a. If $\sum_{i=1}^{n} c_m(O_{mi}) > c_m(O_m)$, the sum of the values of the characteristic c_m after the decomposition of the things is greater than that of the original things O_m about c_m;

b. If $\sum_{i=1}^{n} c_m(O_{mi}) = c_m(O_m)$, the sum of the values of the characteristic c_m after the decomposition of the things is equal to that of the original things O_m about c_m; and

c. If $\sum_{i=1}^{n} c_m(O_{mi}) < c_m(O_m)$, the sum of the values of the characteristic c_m after the decomposition of the things is less than that of the original things O_m about c_m.

Opening-up rule 3.4.3 (Expandability/contractibility rule) Any basic-element can be expanded or contracted under certain conditions. Namely, suppose $B = (O, c, v)$. Under certain condition ℓ, there must be a real number $\alpha(\alpha > 0)$ such that $\alpha B = (\alpha O, c, \alpha v)$. When $0 < \alpha < 1$, the basic-element B can be contracted to αB; when $\alpha > 1$, the basic-element B can be expanded to αB, where αO represents an object with a value of αv.

Case Analysis:

Example 3.4.1

If the illumination function of a desk lamp and the temperature measurement function of a thermometer are combined, we can obtain a multifunctional lamp

D' that can measure the room temperature. According to Opening-up rule 3.4.1, we have:

$$
\begin{bmatrix}
\text{Light up,} & \text{dominating object,} & \text{room} \\
 & \text{tool,} & \text{table lamp } D_1
\end{bmatrix}
$$
$$
\oplus
\begin{bmatrix}
\text{Measuring,} & \text{dominating object,} & \text{temperature} \\
 & \text{tool,} & \text{thermometer } D_2
\end{bmatrix}
$$
$$
=
\begin{bmatrix}
\text{Light up} \oplus \text{Measuring,} & \text{dominating object,} & \text{room} \oplus \text{temperature} \\
 & \text{tool,} & \text{table lamp } D'
\end{bmatrix}
$$

where we have table lamp D' = table lamp $D_1 \oplus$ thermometer D_2.

Example 3.4.2

Suppose

$$
M_1 = \left(\text{Wall lamp } D_1, \quad \text{cost,} \quad 100 \text{ yuan} \right) = \left(O_{m1}, \quad c_{m1}, \quad c_{m1}(O_{m1}) \right).
$$

According to Opening-up rule 3.4.1, at least one matter-element

$$
M_2 = \left(\text{Lampshade } D_2, \quad \text{beautiful degree,} \quad \text{good} \right) = \left(O_{m2}, \quad c_{m2}, \quad c_{m2}(O_{m2}) \right)
$$

can be found, satisfying

$$
M_1 \oplus M_2 =
\begin{bmatrix}
O_{m1} \oplus O_{m2}, & c_{m1}, & c_{m1}(O_{m1}) \oplus c_{m1}(O_{m2}) \\
 & c_{m2}, & c_{m2}(O_{m1}) \oplus c_{m2}(O_{m2})
\end{bmatrix}.
$$

If $c_{m1}(O_{m2}) = 50$ yuan, $c_{m2}(O_{m1})$ = ordinary, then

$$
M_1 \oplus M_2 =
\begin{bmatrix}
\text{Wall lamp } D_1 \oplus \text{Lampshade } D_2, & \text{cost,} & 100 \text{ yuan} \oplus 50 \text{ yuan} \\
 & \text{beautiful degree,} & \text{ordinary} \oplus \text{good}
\end{bmatrix}
$$

Example 3.4.3

Suppose that $M_1 =$ (Light tube D_1, power, 40 W) and $M_2 =$ (Lamp holder D_2, length, 1 m are given, then

$$
M_1 \oplus M_2 =
\begin{bmatrix}
\text{Light tube } D_1 \oplus \text{Lamp holder } D_2, & \text{power,} & 40 \text{ W} \oplus \phi \\
 & \text{length,} & a \text{ m} \oplus 1 \text{ m}
\end{bmatrix}
$$

illustrates that a 40 W power, am long light tube D_1 is aggregated with a 1-m long lamp holder D_2. Whether the two matter-elements can form a system depends on the value of a. Generally speaking, for a light tube and a lamp holder to match, they must satisfy certain requirements. When this is the case, M_1 and M_2 form a system:

$$M_1 \otimes M_2 = \begin{bmatrix} \text{Light tube } D_1 \otimes \text{Lamp holder } D_2, & \text{power,} & 40 \text{ W} \\ & \text{length,} & a \text{ m} \otimes 1 \text{ m} \end{bmatrix} (a < 1).$$

Example 3.4.4

Suppose that a bundle of chopsticks O_m is made up of 10 chopsticks, $O_{m1}, O_{m2}, \ldots, O_{m10}$, namely $O_m = O_{m1} \otimes O_{m2} \otimes \cdots \otimes O_{m10}$. If c_1 is the strength characteristic and we denote $M_1 = (O_m, \ c_1, \ v_1), M_{1i} = (O_{mi}, \ c_1, \ v_{1i}), i = 1, 2, \ldots, 10$, then we have

$$v_1 > \sum_{i=1}^{10} v_{1i}.$$

Suppose c_2 is the weight characteristic. If we denote $M_2 = (O_m, c_2, v_2)$, $M_{2i} = (O_{mi}, \ c_2, \ v_{2i}), \ i = 1, 2, \ldots, 10$, then we have

$$v_2 = \sum_{i=1}^{10} v_{2i}.$$

Suppose c_3 represents the length characteristic. If we denote $M_3 = (O_m, \ c_3, \ v_3)$, $M_{3i} = (O_{mi}, \ c_3, \ v_{3i}), i = 1, 2, \ldots, 10$, then we have

$$v_3 < \sum_{i=1}^{10} v_{3i}.$$

Example 3.4.5

By using the decomposition control of chandeliers on light, we can form different patterns, thus decorating the room. According to Opening-up rule 3.4.2, by letting t be the time parameter we have

$$\left(\text{Chandelier } D(t), \quad \text{light pattern,} \quad \text{rhombus} \right)$$

$$// \left\{ \left(\text{Chandelier } D(t_1), \quad \text{light pattern,} \quad \text{rhombus} \right), \left(\text{Chandelier } D(t_2), \quad \text{light pattern,} \quad \text{round} \right), \right.$$

$$\left. \left(\text{Chandelier } D(t_3), \quad \text{light pattern,} \quad \text{square} \right), \left(\text{Chandelier } D(t_4), \quad \text{light pattern,} \quad \text{mixed form} \right) \right\}$$

Example 3.4.6

By making the irradiation angle of a wall lamp larger or smaller, we can form different patterns, thus decorating the room.

Suppose $M = $ (Wall lamp D, irradiation angle, $30°$). According to Opening-up rule 3.4.3, the irradiation angle of the wall lamp can be enlarged as much as four times, namely $4M = $ (Wall lamp D', irradiation angle, $120°$), forming a new style wall lamp.

Basic Steps of the Decomposition/Combination Chain Method

The decomposition/combination chain method is developed based on the above opening-up rule, using domain knowledge to judge the possibility of whether the basic-elements can be combined, decomposed, expanded, or contracted to find additional ways to innovate or solve contradictory problems. Combination, decomposition, expansion, and contraction are all effective means to innovate or solve contradictory problems.

The steps of the decomposition/combination chain method consist of:

1. Express the object to be analyzed as a basic-element B;
2. Expand the basic-element B using the divergence tree method, leading to the development of several basic-elements;
3. Judge whether B can be combined with other basic-element(s) or can be decomposed, expanded, or contracted according to the domain knowledge; and
4. Investigate whether the basic-element after combination, decomposition, expansion, or contraction can be used to innovate or solve contradictory problems.

Case Analysis:

Example 3.4.7

It is difficult to measure the thickness of a thin sheet of paper by using a common ruler with a measuring range of $[0,100]$ mm. The matter-element of this situation is:

$$M = (\text{Paper } O_{m1}, \quad \text{thickness}, \quad x \text{ mm}), \ x \ll 1 \text{ mm}.$$

According to the paper composability (apparently it cannot be decomposed), we can find M's composable matter-elements as

$$M_i = (\text{Paper } O_{mi}, \quad \text{thickness}, \quad x \text{ mm}), i = 2, 3, \ldots, 100.$$

The goal of the original problem becomes the measurement of the following matter-element:

$$M' = \sum_{i=1}^{100} (O_{mi}, \quad \text{thickness}, \quad x \text{ mm}) = \left(\sum_{i=1}^{100} O_{mi}, \quad \text{thickness}, \quad 100x \text{ mm} \right).$$

Obviously, $100x \in [1,100]$. Namely the thickness of 100 sheets can be measured. Assuming it is measured as 20 mm, the original problem can be solved: $100x = 20$, and thus $x = 0.2$ mm.

Example 3.4.8

On the ceiling of an office is a ceiling lamp, which is 3.2 m from the ground. The lamp tube is broken and needs to be replaced, but there is no ladder. Someone is only 1.75 m tall and can touch a height of 2.25 m. Suppose $M =$ (people D, height, 1.75 m). Suppose that the person's touch-height is 0.5 m taller than his height. So he can provide a height for repairing the lamp: $c_{0t}(M) = 2.25$ m.

The following is a path to solving this contradictory problem using extensible analysis methods:

Using the divergence tree method:

$$M = (\text{People } D, \text{ height}, 1.75\,\text{m}) \dashv \begin{cases} M_1 = \left(\text{People } D_1, \text{ height}, 1.60\,\text{m}\right) \\ M_2 = \left(\text{People } D_2, \text{ height}, 1.85\,\text{m}\right) \\ M_3 = \left(\text{People } D_3, \text{ height}, 1.95\,\text{m}\right) \\ M_4 = \left(\text{Desk } D_4, \text{ height}, 1.10\,\text{m}\right) \\ M_5 = \left(\text{Chair } D_5, \text{ height}, 0.60\,\text{m}\right) \\ M_6 = \left(\text{Cabinet } D_6, \text{ height}, 1.00\,\text{m}\right) \end{cases}$$

And for the tallest person D_3, we have $c_{0t}(M_3) = 1.95 + 0.5 = 2.45$(m) < 3.2(m). Therefore, the first three matter-elements cannot solve the contradictory problem.

Using the decomposition/combination chain method:

$$M \oplus M_1 = (\text{People } D, \text{ height}, 1.75\text{ m}) \oplus (\text{People } D_1, \text{ height}, 1.60\text{ m})$$
$$M \oplus M_4 = (\text{People } D, \text{ height}, 1.75\text{ m}) \oplus (\text{Desk } D_4, \text{ height}, 1.10\text{ m})$$
$$M \oplus M_5 = (\text{People } D, \text{ height}, 1.75\text{ m}) \oplus (\text{Chair } D_5, \text{ height}, 0.60\text{ m})$$
$$M \oplus M_6 = (\text{People } D, \text{ height}, 1.75\text{ m}) \oplus (\text{Cabinet } D_6, \text{ height}, 1.30\text{ m})$$

we have

$$c_{0t}(M \oplus M_1) = 1.75 + 0.8 + 0.5 = 3.05\text{(m)},$$

Namely, person D_1 sits on the shoulders of person D. In this case, person D_1's touch height is about half of his height plus 0.5 m:

$c_{0t}(M \oplus M_4) = 1.75 + 0.5 + 1.10 = 3.35$(m), namely, person D stands on the table D_4;
$c_{0t}(M \oplus M_5) = 1.75 + 0.5 + 0.60 = 2.85$(m), namely, person D stands on the chair D_5;
$c_{0t}(M \oplus M_6) = 1.75 + 0.5 + 1.30 = 3.55$(m), namely, person D stands on the cabinet D_6.

Thus, a variety of possible paths to solving contradictory problems are obtained.

Example 3.4.9

A customer orders 10,000 units of product D from factory E for delivery 15 days later. The factory producing the product has two production lines. Each line only produces 500 units of product D every day. According to the original plan, the production task can be completed 5 days ahead of the delivery date, but after signing the contract, the factory suddenly discovered that one of the production lines suffers from a serious problem, which cannot be resolved in a short period of time. Because one production line can only produce 7,500 units of the product in 15 days, there will be a shortage of 2,500 units. How can the problem be resolved?

The goal affair-element of the problem is

$$
A_g = \begin{bmatrix}
\text{Produce,} & \text{dominant object,} & \text{product } D \\
& \text{acting object,} & \text{factory } E \\
& \text{quantity,} & 10000 \\
& \text{time,} & 15 \text{ days}
\end{bmatrix},
$$

Obviously, under the present condition, the factory cannot complete the production task within 15 days. Namely, A_g cannot be achieved.

According to the divergence tree method, Divergence rule 3.1.6 implies

$$
A_g - | \begin{cases} A_1 \\ A_2 \end{cases}
$$

where

$$
A_1 = \begin{bmatrix}
\text{Produce,} & \text{dominating object,} & \text{product } D \\
& \text{acting object,} & \text{factory } E \\
& \text{quantity,} & 7500 \\
& \text{time,} & 15 \text{ days}
\end{bmatrix},
$$

$$
A_2 = \begin{bmatrix}
\text{Produce,} & \text{dominating object,} & \text{product } D \\
& \text{acting object,} & \text{factory } F \\
& \text{quantity,} & 2500 \\
& \text{time,} & 10 \text{ days}
\end{bmatrix},
$$

Other than factory *E,* there are other factories that can produce this product. According to Opening-up rule 3.4.1, we have:

$$
A_1 \oplus A_2 = \begin{bmatrix} \text{Produce,} & \text{dominating object,} & \text{product } D \\ & \text{acting object,} & \text{factory } E \oplus \text{factory } F \\ & \text{quantity,} & 10000 \\ & \text{time,} & 15 \text{ days} \oplus 10 \text{ days} \end{bmatrix},
$$

Thus, a solution to the contradictory problem can be obtained: Ask an alternative factory *F* to produce 2,500 units of product *D*. This factory *F* can complete producing task in 10 days, thus solving the problem.

Thinking and Exercises

1. From an existing table lamp, write a multi-dimensional matter-element, and use a divergence tree to analyze it.
2. From the function of an existing desk lamp, write a multi-dimensional affair-element, and use a divergence tree to analyze it.
3. From the relationship between the parts of an existing desk lamp, write a multi-dimensional relation element, and use a divergence tree to analyze it.
4. According to the domain knowledge or common sense, find out the correlation network of a pen. Consider the possibility of compulsively establishing a relationship or eliminating a relationship. Is it possible to get new product ideas based on this analysis?
5. According to the domain knowledge or common sense, find out the correlation network of a laser pen. Consider the possibility of compulsively establishing a relationship or eliminating a relationship. Is it possible to get new product ideas based on this analysis?
6. According to the domain knowledge or common sense, find out the correlation network of a lamp. Consider the possibility of compulsively establishing a relationship or eliminating a relationship. Is it possible to get new product ideas based on this analysis?
7. Can a mobile phone be decomposed into the standardized modulation structure? If different functions are put on different modules so that the mobile phone can be DIY (do it yourself), then consumers will have different experiences. Use extensible analysis methods to develop this product idea from existing mobile phones and explain what kind of extensible methods are applied.
8. Can a pen be made into a DIY structure? Can a pen also be a neutral pen, and a pencil? Use extensible analysis methods to develop this product idea from existing pens and explain what kind of extensible methods are applied.

Chapter 4

Basis for Generating Creative Ideas (2)
Conjugate Analysis Method

Content Summary

Studying matter's structure helps us to solve contradictory problems by looking at various parts of the matter. Systems theory studies the matter through the components of the matter as a system and internal and external relations, which is a description of the matter's structure. Through an analysis of the matter, we find that besides systemic properties, the structure of the matter can be studied from its material, dynamic and antithetic properties. The corresponding method of analysis is called conjugate pair method.

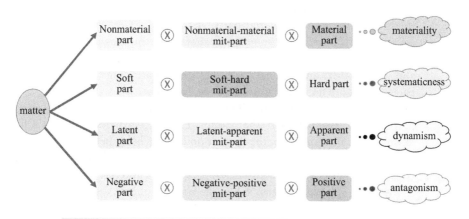

4.1 Conjugation and Conjugate Rules of Matters

QUESTIONS AND THINKING:

- Why can Zhuge Liang's empty-fort strategy make Sima Yi's 100,000 armies retreat?
- Why can advertising increase sales?
- For the statement that "Three stooges are better than one Zhuge Liang" and the statement that "Three monks have no water to drink," what are their respective mechanisms?
- What's the point of the claim that "misfortune generates happiness, and happiness breeds misfortune"?
- Why can people "make use of waste" and "recycle waste"?

4.1.1 Matter's Conjugate Parts

Understanding a matter through its material, systemic, dynamic, and antithetic properties can acquaint people with the matter's structure more completely and more profoundly, revealing the nature of development and change. Therefore, we have proposed the nonmaterial and material, soft and hard, latent and apparent, negative and positive—four pairs of concepts to describe the composition of the matter of concern, called the conjugate parts of the matter.

1. **Nonmaterial part, material part, and nonmaterial and material mit-part**
   In view of materiality, any matter is composed of a material part and a nonmaterial part. We regard the physical part of the matter as the material part, and the non-physical part the nonmaterial part of the matter.

   The material is a foundation, the nonmaterial relates to application. The combination of material and nonmaterial parts constitutes the matter.

   For example, the physical parts of a house, such as the walls, ceilings, and floors are the material parts. Because we are living and working in the space, these activities are nonmaterial parts. A cell phone's entity is the material parts while its brand, appearance, software and so on are nonmaterial parts. For a processing enterprise, capital, equipment, plant, products, personnel, and other entities are material parts, while corporate image, technology, and patent rights are nonmaterial parts of the enterprise.

   In addition, we regard the intermediary part between the nonmaterial part and the material part the "nonmaterial and material mit-part". For example, for an empty cup, the space it potentially holds belongs to the nonmaterial part of the cup. When the cup holds some water, the cup has a material part (the water), and a nonmaterial part (space), and the two parts change as the water volume changes. When the water is poured out, the space part becomes a nonmaterial part. Here, the space containing water is called the nonmaterial and material mit-part.

2. **Soft part, hard part, and soft and hard mid-part** In view of a matter's systemic property, we regard entire set of all parts constituting the matter as the hard part, while the set of relations the matter makes up the soft part of the matter.

   A combination of the soft part and the hard part leads to the matter. The soft part is valuable. It includes the inner-subordinate relationship, the outer-subordinate relationship, the outer-associative relationship, and so on.

   There are two Chinese sayings: "A monk carries water to drink, two monks raise up the water to drink, and three monks have no water to drink"; and "Three stooges are more than one Zhuge Liang." The message of both sayings is that the power of three people depends significantly on whether and how well they are working together. Therefore, it is not enough to merely study the components of the matter to understand the matter and its functionality. We should also study its internal and external relations. For a malfunctioning machine, maybe each of its parts is intact, but because of a broken connection between the parts it doesn't work properly. If we do not pay attention to the soft part, even if every part has been inspected, the cause of the failure cannot be found.

   In addition, if a part of the matter connects another two components together, what is involved is the hard part of the matter as well as the soft part of the matter. To facilitate the systemic analysis of the matter, these parts are regarded as the "soft and hard mit-part" of the matter. For example, all of the wires connecting a computer host, a monitor, and a printer are called the "computer's soft and hard mit-part."

3. **Latent part, apparent part, and latent and apparent mit-part** In view of the matter's dynamic property, it is constantly changing. Being static is always relative, while changing is eternal. A disease has a latent process; and a growing seed has to go through a germinating process. Eggs placed in a certain temperature over a certain period of time will hatch into chickens. We call a matter's potential part the latent part, and the explicit part is the apparent part.

   People and matters all have latent parts and apparent parts. A combination of the latent part and the apparent part leads to the appearance of a matter.

   The latent parts of some matters will become apparent under certain conditions, such as the fetus in a pregnant woman (a latent person) will become an infant (an apparent person). Some latent parts may not become apparent under certain conditions. For example, seeds in the absence of water will not germinate. The latent part of some matters may be nonexistent, equal to the empty set. Some of matters' apparent parts may have latent functions or latent characteristics. For example, the latent function of some beverage bottles is to hold liquid; the un-started air conditioner has potential power consumption. The apparent part of some matters may be latently dangerous. For example, laptop batteries if the temperature is too high may cause the laptop to catch fire or explode. The latent part of some matters may have an apparent function or an apparent characteristic. For example, the fetus can move in the mother's body, absorb nourishment, and have weight, etc.

There must be a critical state in the mutual conversion between the latent part and the apparent part. This critical state is called the "latent and apparent mit-part," such as a baby bird that is breaking the shell, the fetus during birth, and so on.

4. **Negative part, positive part, and negative and positive mit-part**
   Considering the antithetic property of the matter, everything has two sides. The matter's antithetic property is based on a certain characteristic. The characteristic value of a matter is the interactive result of the positive part that produces positive values and the negative part that produces negative values. The part of a matter that takes a positive value about a characteristic is called the positive part of the matter about that characteristic, and the part of the matter that takes a negative value about a characteristic is called the negative part of the matter about that characteristic.

   In addition, between the negative and the positive part, there is also a part of which the value about a characteristic is 0, such as an organization with a balance of income and expense, at which point there is zero profit. The part of the matter that takes 0 about a characteristic is called the "negative and positive mit-part" about the characteristic.

Note: There is a difference between "positive and negative" and "pros and cons." For example, in terms of an enterprise's profit, waste water, waste gas, and waste residue need to be processed. The value of these parts with regard to profit is negative, so it is the negative part of the enterprise. Because these "three wastes" will cause environmental pollution, they are seen as the enterprise's "malpractice." In terms of the enterprise's profit, the employee welfare department, child care benefit, public relations department, and so on, of which the values with regard to profit are negative, are considered some the enterprise's negative parts. These parts, however, potentially promote the working passion of the staff and enhance the enterprise image, and thus they are "advantageous" parts of the enterprise. In other words, the negative part about a characteristic may be a beneficial part of the matter, and it may also be a passive part of the matter.

4.1.2 Conjugate Rules

It is known from these conjugate properties of the matter that conjugate analysis of the matter includes the nonmaterial and material conjugate analysis, soft and hard conjugate analysis, negative and positive conjugate analysis, and latent and apparent conjugate analysis. Moreover, the matters involved in the resolution of contradictory problems, no matter whether they are subjective or objective or resources, should obey the following conjugate analysis rules.

Conjugate rule 4.1.1 Any matter has conjugate parts such that the sum of each pair of conjugate parts and the corresponding mit-part is equal to the original matter. Namely, suppose that the matter is O_m, and thus its material part is $re(O_m)$, the non-material part is $im(O_m)$, the nonmaterial and material mit-part is $mid_{re\text{-}im}(O_m)$, the soft part is $sf(O_m)$, the hard part is $hr(O_m)$, the soft and hard mit-part is $mid_{sf\text{-}hr}(O_m)$, the latent part is $lt(O_m)$, the apparent part is $ap(O_m)$, the latent and apparent mit-part is $mid_{lt\text{-}ap}(O_m)$, the negative part about characteristic c is $ng_c(O_m)$, the positive part about characteristic c is $ps_c(O_m)$, the negative and positive mit-part is $mid_{ng\text{-}ps}(O_m)$; then the following holds true:

$$O_m = re(O_m) \otimes im(O_m) \otimes mid_{re\text{-}im}(O_m)$$

$$= hr(O_m) \otimes sf(O_m) \otimes mid_{sf\text{-}hr}(O_m)$$

$$= lt(O_m) \otimes ap(O_m) \otimes mid_{lt\text{-}ap}(O_m)$$

$$= ng_c(O_m) \otimes ps_c(O_m) \otimes mid_{ng\text{-}ps}(O_m).$$

Usually, due to the limitation of human cognitive ability, the mit-part between each pair of conjugate parts is not studied alone, but often combined with a conjugate part for discussion according to the practical problems under consideration.

For some matter, one of its conjugate parts may be non-existent or empty, and its mit-part may also be non-existent or empty. For a music CD, the disc is the material part, while the music in it is the nonmaterial part. The so-called "empty disc," in fact, is the disc with an empty nonmaterial part. For dead ancestors, although their bodies no longer physically exist, their reputation, spirit, and other nonmaterial parts still exist.

Conjugate rule 4.1.2 In a pair of conjugate parts of any matter, at least one characteristic of a conjugate part is correlative with a characteristic of its corresponding conjugate part. Namely:

1. Any matter has a nonmaterial part and a material part. At least one characteristic of the nonmaterial part is correlative with a characteristic of the material part;
2. Any matter has a soft part and a hard part. At least one characteristic of the soft part is correlative with a characteristic of the hard part;
3. Any matter has a negative part and a positive part. At least one characteristic of the negative part is correlative with a characteristic of the positive part; and
4. Any matter has a latent part and an apparent part. At least one characteristic of the latent part is correlative with a characteristic of the apparent part.

Therefore, to comprehensively analyze the matter, we must analyze it through its conjugate parts—not only the composition of the conjugate parts, but also correlations between each pair of conjugate parts. Only in this way can the mistakes of "taking a part as the whole" and "attending to one aspect and losing another" be avoided.

4.2 Conjugate Pair Methods

Using the conjugate rules to analyze the matter of concern comprehensively and to solve the given contradictory problem is called conjugate analysis. Because the conjugate analysis is the analysis of the object of concern by way of the nonmaterial and material pair, the soft and hard pair, the latent and apparent pair, and the negative and positive pair, the method is also called the conjugate pair method.

The conjugate pair methods include the nonmaterial and material conjugate pair method, the soft and hard conjugate pair method, the latent and apparent conjugate pair method, and the negative and positive conjugate pair method. The conjugate pair methods are mainly used to analyze matters, people, resources, needs, and requirements, and to perform conjugate classification on them.

4.2.1 Conjugate Classification of Products

When using the conjugate pair method to analyze products, we can divide the products under consideration into material products and nonmaterial products, soft products and hard products, latent products and apparent products, and negative products and positive products.

1. **Material products and nonmaterial products** Products using a material entity to satisfy people's needs are known as material products. For example, food, clothes, housing, furniture, and so on, are material products.

   Products meeting people's needs in a nonmaterial form are called nonmaterial products. Products such as music works, dance works, movies, books, software, and so on, are nonmaterial products. Nonmaterial products often use material products as their carriers. For example, software products use CDs, mobile hard disks, etc., as their carriers. Words and pictures in books use paper as their carrier.

2. **Hard products and soft products** The products that meet people's needs by constructing the hard parts of a system are hard products. For example, a computer mainframe, a monitor, speakers, a mouse, a keyboard, etc., are the hard products of the computer system. Products that meet people's needs through establishing certain relationships are called soft products. For example, the connecting wires and screws of the computer's components are soft products of the computer system.

3. **Apparent products and latent products** A product that can immediately meet the needs of people is called an apparent product. A product with a

certain potential function is called latent product. This latent function actually exists objectively, but further understanding and development are still needed to make it apparent. For example, the packaging box in which a child's toy is sold can be used as a stool for children to sit. In this manner, the box is a latent product.

4. **Positive products and negative products** For a particular characteristic, the products that are beneficial to people are called positive products about this characteristic, and products that are disadvantageous to people are called negative products. For people's health, food, beverages, etc., are positive products, while arsenic is a negative product. In the treatment of a disease, however, arsenic may be helpful. In this case, it is a positive product. That is, positive and negative products are distinguished by whether people's needs are satisfied by the products' functions.

4.2.2 Conjugate Classification of Resources

1. **Nonmaterial and material conjugate resources** In view of materiality of resources, resources may be categorized into material resources and non-material resources. Physical resources are called material resources, and non-physical resources are called nonmaterial resources. For example, an enterprise's personnel, capital, workshop, equipment, land and other tangible assets are material resources; the enterprise's staff's intelligence, knowledge, reputation, brand, patent, information, and even time and space, etc., are nonmaterial resources.

   Most enterprises pay sufficient attention to material resources. In recent years, many enterprises have established a relatively deep acquaintance with some nonmaterial resources. For example, the brand, patent, information, intellectual resources, and so on, have each been exclusively studied. Although it has long been proposed that "time is money," there are few studies about the use of time and space resources.

   In addition, with the continuous progress of technology, the advent of the internet era, "attention," "clicks," and "praise," which have never been taken seriously, are also considered to be very important nonmaterial resources.

2. **Soft and hard conjugate resources** In view of systematic properties of resources, resources can be categorized into hard resources and soft resources. The parts making up resources are called hard resources. Various relationships between resources are called soft resources.

   The hard resources of an enterprise include the components of funds, workshop, equipment, personnel, and others that are not detailed here. The soft resources of the enterprise include its various internal and external relationships. In other words, soft resources are the relationship resources of the enterprise. The richer the soft resources of the enterprise, the higher the quality and the higher the success rate the enterprise enjoys.

We should pay more attention to developing and utilizing soft resources. Sometimes soft resources become the key to the success of the enterprise. Planning activities of the enterprise involve the use of relationship resources. When selling products, for example, we should consider the relationship with customers, the commercial sector, the tax sector, and so on. When financing, we should consider the relationship with the financial sector. When advertising, we should consider the relationship with the media and with policy and regulation bodies. In larger activities, we should consider the relationship with governmental agencies and make efforts to get the support of the government. All of these relations are the soft resources of the enterprise. Soft resources will convert into economic resources of the enterprise under certain conditions.

The majority of the enterprise's soft resources has a low degree of being controlled and may even be uncontrollable. If we want to use soft resources, we must study how to make it as easy as possible to control them.

3. **Negative and positive conjugate resources** In view of antithetic properties, resources can be categorized into positive resources and negative resources. So-called positive resources are those that can promote the development of the enterprise; negative resources are those that hinder the development of the enterprise.

   Because positive resources are self-explanatory, we mainly analyze negative resources. For example, the laid-off workers of an enterprise, who are still looking for jobs, to a certain degree and over a certain period of time, are the enterprise's negative resources. Waste materials, waste water, waste residues, and so on, produced in the production process of an enterprise are also negative resources. An adverse relationship resource, such as a deteriorating internal or external relationship with someone in an enterprise, a bad reputation in the market caused by improper handling of a certain problem, etc., are negative resources of the enterprise. If we can exploit and utilize them appropriately, we can turn negative resources into positive ones. For example, "renewable resources" are developed for negative resources; what can become a positive resourcs is called a renewable resource. For example, some enterprises use scrap to make delicate, stylish products, use waste water as workers' heating water, and use waste residues to make bricks, pave roads, and so on.

   Besides using positive resources, enterprises should pay special attention to the conversion and utilization of negative resources. Some negative resources must be converted, such as illegal negative resources and harmful relationship resources.

4. **Latent and apparent conjugate resources** In view of dynamic properties of resources, resources can be categorized into apparent resources and latent resources. Apparent resources are obvious and can be used directly by enterprises. Latent resources either don't exist explicitly or, if they obviously exist, they cannot be directly used. With the passing of time and with a certain transformation, latent resources can become available.

For example, the relationship between an enterprise and the financial sector when no loan is needed is a latent resource of the enterprise. Once financing is needed, the enterprise can take full advantage of this resource to obtain the capital necessary for the operation of the enterprise, thus making it apparent. A heavily indebted enterprise, when it goes bankrupt, will not be totally worthless because its workshops, equipment, even brands, patents, etc., are potential resources. Once combined with other enterprises and injected with a certain amount of capital, these resources will become immediately available, thus making these latent resources apparent.

Other examples of latent resources include undeveloped oil resources, undeveloped intellectual resources, unused space resources, enterprises idle workshop, equipment, etc.

4.2.3 A Product and Its Conjugate Analysis

The product is a very common and very old concept; its connotation and extension are in a continuous process of development and change. Products are the core starting point of all activities of each and every enterprise. Enterprises want to occupy a favorable position in the fierce market competition, and for this to happen the most important thing is to have products favored by consumers. Otherwise, all the activities of the enterprise will be fruitless.

1. **The concept and structure hierarchy of a product** The traditional concept of a product maintains that the product is an entity with a specific material form and specific practical uses, such as television sets, all kinds of furniture, clothes, and so on. However, with the gradual development and improvement of the market, the concept has broken the limitations of traditional thinking, and products are no longer limited to people's visual, auditory, and other perceptual aspects. In modern marketing theories, the definition of a "product" differs in thousands of different ways, but the essence is the same. Modern marketing theories hold that a product generally includes three levels: the core layer, the formation layer, and the extension layer. Figure 4.2.1 shows the essential parts of a product.

2. **Conjugate analysis of a product** According to the conjugate property of matter, any product has a nonmaterial part and a material part, a soft part and a hard part, a latent part and an apparent part, a negative part and a positive part. All of these parts affect each other and convert into each other. In a product analysis, to fully understand the product, we must start from these four angles.

   a. **Nonmaterial part and material part of a product.** The physical part of the product is the material part of the product, and its non-physical part is the nonmaterial part of the product. Taking a color television as an

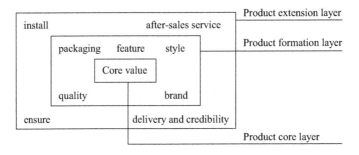

Figure 4.2.1   The structure hierarchy of a product.

example, the material part, nonmaterial part, and corresponding characteristics and values are as shown in Table 4.2.1:

When the homogeneity degree of some product satisfying a particular need is excessively high in the market, the competition of the products inevitably turns to the competition of the nonmaterial parts of the products. Accordingly, the marketing efforts are also turned to the nonmaterial parts of the products. As a famous beer manufacturer once said: "We don't sell beer, we sell hope." For the substitutable products from different factories, because their brands are different and their reputations are

Table 4.2.1   Material Part and Nonmaterial Part of a Product

Product	Conjugate Part	Characteristic	Value
Television set *D*	material part	quality	excellent
		cost	3,000 yuan
		service life	8 years
		color	colors
		shell material	all plastics
		shape	plane right angle
		shell color	black
		weight	30 kg
		model	JC-186
		size	29 inches
		......	......
	nonmaterial part	brand name	a
		trademark name	b
		logo	c
		brand value	d
		popularity	e
		credibility	f
		reputation	g
		......	......

different, even if the qualities of the real parts are the same, customers have different purchase interests in them, thus the market shares are also quite different for the substitutable products.

b. **Soft part and hard part of a product.** The components of the product, such as all parts, accessories, packaging, etc., are the hard part of the product. The relationships the product involves (such as the whole of the internal relationship and the external relationship) are called the soft part of the product. For a television set, the internal connection relations between various parts of the television, the connection between the television and the external power supply, video recorder, VCD, DVD, speakers and other connection relations, after-sales service (which reflect the relations between TV and customers, manufacturers, distributors and service providers), etc., are collectively known as the soft part of the television set.

Services in the soft part of the product (including pre-sale, sale, after-sales service) are the specific forms of the relationship between the product and enterprises, the relationship between the product and customers, and the relationship between the product and distributors. The construction of the soft part of the product is extremely important. If the product's structure is unreasonable, inconvenient to use, or in the design process it did not take into account its effect on other related products, then the product will not be conducive to its sale, leading to an uneasy experience in the marketplace. If the service is not good enough, huge losses may be caused for the enterprise.

c. **Latent part and apparent part of a product.** Each product has many obvious functions, including use functions and auxiliary functions. Most of these functions are clearly required in the design phase of the product. But because a product has many functions, the product can also have many potential functions. Under certain conditions, these potential functions become apparent.

For example, the drug aspirin, which everyone is very familiar with, has the obvious function of treating pain and fevers, but it also has a potential function of reducing the viscosity of the blood. Before this function was found, people just regarded it as a medicine to relieve pain and fever. After this new function was found (apparent), the medicine has been widely used for the prevention and treatment of cardiovascular and cerebrovascular diseases, and its capability to reduce pain has become much less significant to customers. Paying attention to potential characteristics of the product and preventing potential dangers are a requirement when creating products.

d. **Negative part and positive part of a product.** A product has multiple characteristics. For a characteristic, it has beneficial parts for people (the positive parts of the product) and harmful parts for people (the negative parts of the product). Paints can beautify a room, which is the positive part of paints. But some paint contains harmful ingredients for human health, which is the negative part of paints (for people's health). Therefore, when creating products, both the positive parts and negative parts should be considered simultaneously.

Case Analysis:

Example 4.2.1

During the Second World War, motels in the United States mainly received frequently transferred military personnel. At the time, the soldiers constituted a special customer group, which drove the motels to be simply equipped, low-serving, and inexpensive, because at the time military personnel mainly wanted to rest and recuperate, and nothing beyond that.

By the early 1950s, however, with the advent of peace and gradual recovery of the economy, these motels' customers had changed from frequently rotated soldiers to merchants and tourists. In addition, there had been a fundamental change in customer demand for hotels:

1. Increased demand for a large number of other "material products." For example, customers expected to get clean and tidy bedding, soap, shampoo, towels, etc.; and
2. The demand for nonmaterial parts of some "material products" and the demand for some "nonmaterial products" became dominant. For example, customers needed televisions to watch programs, telephones for communication, air-conditioning to adjust the room temperature, flowers to make the environment pleasant, express checkout services, and a quiet and elegant environment.

Facing these new requirements of the customers, most motel and hotel operators turned a blind eye and stuck to the convention ... and they saw their operating profits decline. But builder Wilson was keen to capture these new demands. After some planning, he created a hotel, Holiday Inn, that could meet these needs. Now, Holiday Inn consists of 1,750 local hotels worldwide and is worth US$3.7 billion. Wilson's success began with the goal of satisfying the needs of the customers, which stemmed from a proper analysis of customer needs.

Example 4.2.2

Last year, website D made a profit of 100,000 yuan. This year, website D has set the goal of "doubling the annual profit." Use the conjugate analysis method of nonmaterial-material resources to analyze how they might achieve this goal.

Suppose the current condition of website D is $L = M_{im} \otimes M_{re}$, where

$$
M_{im} = \begin{bmatrix}
\text{Website } D, & \text{number of clicks per day,} & 1{,}000 \\
 & \text{popularity,} & 1 \\
 & \text{grade of work,} & 2 \\
 & \text{software innovation degree,} & 1
\end{bmatrix}
=
\begin{bmatrix}
D, & c_1, & 1{,}000 \\
 & c_2, & 1 \\
 & c_3, & 2 \\
 & c_4, & 1
\end{bmatrix},
$$

$$M_{re} = \begin{bmatrix} \text{Website } D, & \text{advertising revenue,} & 300,000 \text{ yuan} \\ & \text{actual annual profit,} & 100,000 \text{ yuan} \end{bmatrix}$$

$$= \begin{bmatrix} D, & c_5, & 300,000 \text{ yuan} \\ & c_6, & 100,000 \text{ yuan} \end{bmatrix}.$$

The goal G is

$$G = \begin{bmatrix} \text{Improve,} & \text{dominating object,} & \text{annual profit} \\ & \text{quantity,} & 200,000 \text{ yuan} \\ & \text{time,} & \text{this year} \end{bmatrix}.$$

Then goal G cannot be achieved under condition L.

Let M_{re} be the material part condition, and M_{im} the nonmaterial part condition. Obviously the problem of the website is that its nonmaterial part condition of the matter element cannot satisfy the need of achieving the material part of the goal. This contradictory problem of the material part is superficial. According to the following correlative analysis

$$\begin{matrix} (D, & c_3, & v_3) \\ (D, & c_4, & v_4) \end{matrix} \Big\} \sim (D, \ c_2, \ v_2) \sim (D, \ c_1, \ v_1)$$

$$\sim (D, \ c_5, \ v_5) \sim (D, \ c_6, \ v_6).$$

Obviously, a change in the rank of the website and the degree of software innovation enhance the site's reputation so that the number of daily clicks increases, which helps promote the increase of advertising and thus the profit.

Its practical significance is to gain more attention through passionate good works and software innovation and thus to get more "clicks" or "visits" (these are nonmaterial resources of the enterprise), so that their site becomes a "well-known site" or a "famous site" on the internet. Thus more profit can be earned through charging enterprises for publishing advertisements or product information, or through online sales (to obtain material resources).

Many internet citizens upload their favorite songs, paintings, articles, and other good things to the internet, which are free to other internet citizens to check out or download. A lot of people volunteer to modify and improve free software like Linux without charge, or even publish the original code. These people's actions are intended to attract public attention, that is, to develop "attention resources" (nonmaterial resources). This is the method of using nonmaterial resources to achieve various goals.

Example 4.2.3

Why could Zhuge Liang's "empty fort strategy" have made Sima Yi's army of over 150,000 soldiers retreat?

In the Three Kingdoms' time period, Zhuge Liang, the prime minister of Shu Kingdom, lost an important strategic location, named Street Pavilion, due to wrongly using the general named Masu. General Sima Yi of Wei Kingdom took advantage of the favorable situation and moved his army of 150,000 soldiers toward Zhuge Liang's western city. At that time, Zhuge Liang had no real armed forces in the city except one class of civil servants and around 5,000 lightly armed guards, half of whom went to transport war supplies away from the city. Facing the sudden appearance of the crisis, Zhuge Liang commanded, "Hide all banners, and soldiers stand still; behead anyone who goes out privately or speaks noisily." And he told the soldiers to open the four city gates, and at each gate around twenty soldiers, dressed as common people, sprinkled water and swept the streets. Zhuge Liang draped himself in a crane cloak and put on a long black silk ribbon scarf. He sat in the front of the watch tower of the city, leaning against a railing, and slowly played a guqin, an ancient Chinese music instrument, with a small child standing on each side and with incense burning. General Sima Yi saw Zhuge Liang comfortably sitting in the front of the tower, smiling, burning incense, and playing his guqin. Sima Yi looked, pondered, and said: "Zhuge Liang has always been cautious, and never took any risk. Now the gates are open with him enjoying the music. So, there must be an ambush planned. If my army goes into the city, we will just fall into his trap. Let's just retreat!" Then his military forces retreated.

In this historical event, Zhuge Liang's material resources are only "2,500 soldiers," which is far inferior to Sima Yi's material resources of "150,000 soldiers." But Zhuge Liang also has the nonmaterial resource of "lifelong cautiousness," which Sima Yi understood well. Therefore, Zhuge Liang used his nonmaterial resources to defeat Sima Yi's material resources. This is an example of a good application of the material-nonmaterial conjugate analysis.

Example 4.2.4

What are the causes of the phenomena that "Three stooges are better than one Zhuge Liang" and that "Three monks have no water to drink"?

Suppose that three stooges form a group, with each of them a part of this group (hard part). Being stooges, each person's decision-making capability is inferior to that of Zhuge Liang. When these stooges solidify, the combined soft part will be very strong. Thus the comprehensive decision-making capability of the group surpasses that of Zhuge Liang.

For the three-monk group, each of them has the ability to carry water (hard part). But because the relationship among them (soft part) is not good, together as a group they do not have the ability to carry the water.

These two phenomena can be analyzed using soft and hard conjugate analysis. This is also a good example of what is often described as "$1 + 1 \neq 2$" in systems science, which can be expressed by the matter-element as follows:

$$\left(\text{Stooge } D_{11}, \quad \text{decision-making level}, \quad v_{11}\right) \otimes \left(\text{Stooge } D_{12}, \quad \text{decision-making level}, \quad v_{12}\right)$$

$$\otimes \left(\text{Stooge } D_{13}, \quad \text{decision-making level}, \quad v_{13}\right) = \left(\text{Stooges } D_1, \quad \text{decision-making level}, \quad v_1\right)$$

and $v_1 > v_{11} \oplus v_{12} \oplus v_{13}$. That is, due to the role of the soft part, the overall capacity is greater than the sum of the parts. Similarly,

$$\left(\text{Monk } D_{21}, \quad \text{water carrying capacity}, \quad v_{21}\right) \otimes \left(\text{Monk } D_{22}, \quad \text{water carrying capacity}, \quad v_{22}\right)$$

$$\otimes \left(\text{Monk } D_{23}, \quad \text{water carrying capacity}, \quad v_{23}\right) = \left(\text{Monks } D_2, \quad \text{water carrying capacity}, \quad v_2\right)$$

and $v_2 < v_{21} \oplus v_{22} \oplus v_{23}$. That is, due to the role of the soft part, the overall capacity is less than the sum of the parts.

There are many other similar examples. For example, a steel rope is composed of many thinner steel wires. Combined, the twisted wires have a higher toughness than the individual thinner steel wires. The steel wires are twisted so that each wire can tolerate the average force, and are not easy to break, thereby increasing the strength of the entire steel rope.

Example 4.2.5

Why do online shops ask for customers to give them praises?

"Praise" was originally a comprehensive evaluation result of goods and online shops. But in view of negative and positive conjugate analysis, in the present internet era, "praise" has become a "positive resource" of the shop. This directly affected the sales of the online shop. When customers go shopping online, they look at the comments of customers who have purchased the products before considering whether they will order from this website. Therefore, "bad comments" become a "negative resource" of the online shop. Online shops often lose a lot of potential customers because of "bad reviews."

Example 4.2.6

Why are faulty cars recalled?

On September 28, 2015, AQSIQ issued an announcement that Mazda (China) Enterprise Management Co., Ltd., starting on September 29, 2015, would recall some of the imported Mazda 6 vehicles (production date is from March 29, 2002 to November 1, 2004), and imported Mazda RX-8 cars (production date is from April 14, 2004 to February 15, 2008). According to the company's statistics, mainland China involved 309 units and 370 units, respectively.

The announcement shared that, for the vehicles within the recall time periods, when the airbags unfold, the gas generators may cause abnormal damage, resulting in debris flying out, which may injure the passengers in the car. Given this safety risk, as a precautionary measure Mazda (China) Enterprise Management Co., Ltd., will replace all airbag gas generators for the vehicles within the recall time periods to reduce the safety risk while carrying out an investigation into the faulty gas generators.

In this case, a car's "airbag gas generator" is its apparent part, while its "safety hazard" is the car's latent part. Without recall and replacement, there may be a risk of danger to the passenger(s).

Thinking and Exercises

1. Why are Baishi Qi's paintings so valuable? Please analyze this question using the conjugate pair method. In a stationery product innovation, is it possible to use Baishi Qi's pictures?

2. Perform a soft and hard conjugate analysis on a table lamp. Consider whether it is possible to obtain an innovative new product idea through changing the soft part.

3. An obvious competitor of Eastman Kodak Company is Fujifilm. Analyze who really defeated Kodak.

4. The situation of China's aging population is serious. What commercial opportunities do you see in this serious situation?

5. Use the conjugate pair method to analyze a laser pointer's imaginary part and real part, soft part and hard part, latent part and apparent part, and negative part and positive part. If one of its negative parts is adverse to people, can you change it to get innovative ideas for new products?

6. Use the conjugate pair method to analyze the imaginary part and real part, soft part and hard part, latent part and apparent part, and negative part and positive part of yourself. Discover your superiority and inferiority, and then propose a method to change the identified inferiority.

Chapter 5

Tools for Generating Creative Ideas
Extension Transformation Methods

Content Summary

Extensible analysis and conjugate analysis introduced in Chapters 3 and 4 can only present a variety of ways to innovate or solve contradictory problems. However, extension transformations must be implemented to derive creative ideas for innovation or solving contradictory problems. Through extension transformations, unknown problems can be turned into known problems, impracticable problems into practicable problems.

The idea of extension transformation is to transform one object into another object or into several objects. The premise of carrying out an extension transformation on a certain research object is the extensible analysis. Transformation paths can be presented with the use of the extensible analysis method.

The most basic extension transformation method consists of five basic transformations: substituting transformation, adding/removing transformation, enlarging/shrinking transformation, decomposing transformation, and duplicating transformation. It has been developed through a great deal of research on creative ideas from all over the world. Furthermore, four operating methods of transformations – conductive transformations, conjugate transformations, and composite transformations—have been developed to form an extension transformation system.

Creative ideas are generated by extension transformations or their operations.

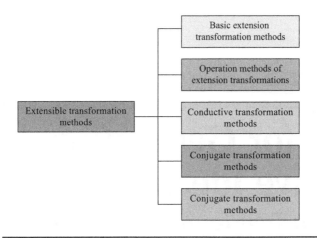

5.1 Basic Extension Transformation Methods

Questions and Thinking:

■ What would you think upon observing Figures 5.1.1 and 5.1.2?
■ What kinds of transformation happens from the left figure to the right one? Take ceiling lamp *D* as an example to illustrate the transformations.

5.1.1 Substituting Transformation of Basic-elements

5.1.1.1 Substituting Transformation on the Values of Basic-elements

The value on a certain characteristic for an object of a certain basic-element is transformed into another value, denoted as $TB = T(O, c, v) = (O', c, v') = B'$.

Illustration

1. Because basic-elements are divergent, there will be numerous objects and values with the same characteristic (Divergence Rule 3.1.2). Thus, a basic-element *B* can be transformed to another basic-element *B'*.

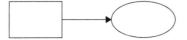

Figure 5.1.1   Schematic diagram of substituting transformation (1).

Figure 5.1.2   Schematic diagram of substituting transformation (2).

2. The substituting transformation of the value of a matter-element will inevitably lead to the conductive transformation of its object, which will be introduced in Section 5.3.
3. The substituting transformation of the value of a certain matter-element or relation-element may not lead to the conductive transformation of its object.

Case Analysis:

Example 5.1.1

The following substituting transformation of matter-elements can be used to express Figure 5.1.1:

$$T_1 \begin{bmatrix} \text{Ceiling lamp } D, & \text{shape,} & \text{cuboid} \\ & \text{color,} & \text{white} \end{bmatrix} = \begin{bmatrix} \text{Ceiling lamp } D', & \text{shape,} & \text{ellipsoid} \\ & \text{color,} & \text{white} \end{bmatrix}.$$

The following substituting transformation of matter-elements can be used to express Figure 5.1.2:

$$T_2 \begin{bmatrix} \text{Ceiling lamp } D, & \text{shape,} & \text{cuboid} \\ & \text{color,} & \text{white} \end{bmatrix} = \begin{bmatrix} \text{Ceiling lamp } D'', & \text{shape,} & \text{cuboid} \\ & \text{color,} & \text{gray} \end{bmatrix}.$$

These symbolic substituting transformations can also be expressed in a table as shown in Table 5.1.1. The subsequent transformation results can also be tabulated.

Example 5.1.2

For Figure 5.1.3, the substituting transformation of the values of "shape" and "color" are applied simultaneously, which can be expressed as follows:

$$T_1 \begin{bmatrix} \text{Ceiling lamp } D, & \text{shape,} & \text{cuboid} \\ & \text{color,} & \text{white} \end{bmatrix} = \begin{bmatrix} \text{Ceiling lamp } D''', & \text{shape,} & \text{ellipsoid} \\ & \text{color,} & \text{gray} \end{bmatrix}.$$

Table 5.1.1 Substituting Transformation for Ceiling Lamp *D*

Before Transformation			Transformation Method	After Transformation		
Object	Characteristic	Value		Object	Characteristic	Value
Ceiling lamp D	shape	cuboid	Substituting transformation	Ceiling lamp D'	shape	ellipsoid
	color	white			color	white
Ceiling lamp D	shape	cuboid	Substituting transformation	Ceiling lamp D''	shape	cuboid
	color	white			color	gray

Figure 5.1.3 Schematic diagram of substituting transformation (3).

Example 5.1.3

The function of "providing light by means of irradiation" of a product with the formalization of affair-elements can be expressed as:

$$A = \begin{bmatrix} \text{Provide,} & \text{dominating object,} & \text{light} \\ & \text{means,} & \text{irradiate} \end{bmatrix}.$$

According to Divergence Rule 3.1.1, multiple matter-elements can be extended by the above functional affair-elements, such as

$$A \dashv \left\{ \begin{bmatrix} \text{Provide,} & \text{dominating object,} & \text{color} \\ & \text{means,} & \text{irradiate} \end{bmatrix} \\ \begin{bmatrix} \text{Provide,} & \text{dominating object,} & \text{music} \\ & \text{means,} & \text{broadcast} \end{bmatrix} \\ \begin{bmatrix} \text{Provide,} & \text{dominating object,} & \text{brightness} \\ & \text{means,} & \text{irradiate} \end{bmatrix} \right\}.$$

The substituting transformation of values can be applied as follows:

$$T_1 \begin{bmatrix} \text{Provide,} & \text{dominating object,} & \text{light} \\ & \text{means,} & \text{irradiate} \end{bmatrix} = \begin{bmatrix} \text{Provide,} & \text{dominating object,} & \text{color} \\ & \text{means,} & \text{irradiate} \end{bmatrix}$$

$$T_2 \begin{bmatrix} \text{Provide,} & \text{dominating object,} & \text{light} \\ & \text{means,} & \text{irradiate} \end{bmatrix} = \begin{bmatrix} \text{Provide,} & \text{dominating object,} & \text{music} \\ & \text{means,} & \text{play} \end{bmatrix}$$

$$T_3 \begin{bmatrix} \text{Provide,} & \text{dominating object,} & \text{light} \\ & \text{means,} & \text{irradiate} \end{bmatrix} = \begin{bmatrix} \text{Provide,} & \text{dominating object,} & \text{brightness} \\ & \text{means,} & \text{irradiate} \end{bmatrix}.$$

Therefore, three affair-elements with new functions for the product can be obtained.

Example 5.1.4

There is an up-down position relation between the lamp holder D_1 and the lampshade D_2 of the wall lamp. Through the formalization of a relation-element, this position relation can be expressed as

$$R = \begin{bmatrix} \text{Up-down position relation,} & \text{antecedent,} & D_2 \\ & \text{consequent,} & D_1 \end{bmatrix}.$$

If the substituting transformation of the following values is applied,

$$TR = T \begin{bmatrix} \text{Up-down position relation,} & \text{antecedent,} & D_2 \\ & \text{consequent,} & D_1 \end{bmatrix}$$

$$= \begin{bmatrix} \text{Up-down position relation,} & \text{antecedent,} & D_1 \\ & \text{consequent,} & D_2 \end{bmatrix} = R',$$

then the new relation-element represents a new relation between the lamp holder D_1 and the lampshade D_2 of the wall lamp, which represents a wall lamp with a new structure where the lamp holder is above the lamp shade.

5.1.1.2 Substituting Transformation of the Objects of Basic-elements

The object of a basic-element is transformed into another object, denoted as:

$$TB = T(O, c, v) = (O', c, v) = B'.$$

At this point, the characteristics and value of the basic-element remain unchanged.

Case Analysis:

Example 5.1.5

There is a lamp in a spherical shape. According to Divergence Rule 3.1.5, the lamp can be extended as follows:

$$\left(\text{Lamp } D, \quad \text{shape,} \quad \text{spherical shape} \right) \dashv \begin{cases} \left(\text{Dining lamp } D_1, \quad \text{shape,} \quad \text{spherical shape} \right) \\ \left(\text{Wall lamp } D_2, \quad \text{shape,} \quad \text{spherical shape} \right) \\ \left(\text{Ceiling lamp } D_3, \quad \text{shape,} \quad \text{spherical shape} \right) \\ \left(\text{Porch lamp } D_4, \quad \text{shape,} \quad \text{spherical shape} \right) \end{cases}.$$

So the substituting transformation of the objects of the matter-element can be applied as follows:

$$T \left(\text{Lamp } D, \quad \text{shape,} \quad \text{spherical shape} \right) = \left(\text{Dining lamp } D_1, \quad \text{shape,} \quad \text{spherical shape} \right).$$

Namely, we have obtained a new idea: a dining lamp in a spherical shape.

Example 5.1.6

With an affair-element, the function "control light" of a product can be expressed as:

$$(\text{Control}, \quad \text{dominating object}, \quad \text{light}).$$

Furthermore, according to Divergence Rule 3.1.5, this affair-element can be expanded to

$$(\text{Control}, \quad \text{dominating object}, \quad \text{light}) \dashv \begin{cases} (\text{Provide}, \quad \text{dominating object}, \quad \text{light}) \\ (\text{Adjust}, \quad \text{dominating object}, \quad \text{light}) \\ (\text{Shield}, \quad \text{dominating object}, \quad \text{light}) \\ (\text{Strengthen}, \quad \text{dominating object}, \quad \text{light}) \\ (\text{Filtrate}, \quad \text{dominating object}, \quad \text{light}) \\ (\text{Absorb}, \quad \text{dominating object}, \quad \text{light}) \end{cases}.$$

So the following substituting transformation of the objects of the function-element can be applied as:

$$T(\text{Control}, \quad \text{dominating object}, \quad \text{light}) = (\text{Adjust}, \quad \text{dominating object}, \quad \text{light}).$$

Namely, light-adjusting, a new creative idea for the product's function can be obtained.

Example 5.1.7

Regarding the up-down position relation between the lampshade D_1 and the lamp holder D_2 of a certain table lamp, it can be expressed with a relation-element as follows:

$$\begin{bmatrix} \text{Up-down position relation}, & \text{antecedent}, & D_1 \\ & \text{consequent}, & D_2 \end{bmatrix}$$

According to Divergence Rule 3.1.5, it can be extended to

$$\begin{bmatrix} \text{Up-down position relation}, & \text{antecedent}, & D_1 \\ & \text{consequent}, & D_2 \end{bmatrix} \dashv \begin{bmatrix} \text{Left-right position relation}, & \text{antecedent}, & D_1 \\ & \text{consequent}, & D_2 \end{bmatrix}.$$

So the following substituting transformation of the objects of the relation-element can be applied:

$$T\begin{bmatrix} \text{Up-down position relation,} & \text{antecedent,} & D_1 \\ & \text{consequent,} & D_2 \end{bmatrix} = \begin{bmatrix} \text{Left-right position relation,} & \text{antecedent,} & D_1 \\ & \text{consequent,} & D_2 \end{bmatrix}.$$

5.1.1.3 Substituting Transformation on Characteristics of Basic-elements

The characteristic of a certain basic-element is transformed into another characteristic, denoted as:

$$TB = T(O, c, v) = (O, c', v') = B'.$$

At this point, the value of the basic-element can either remain unchanged or become a new value.

Case Analysis:

Example 5.1.8

The height and width of a wall lamp D are 0.5 m and 0.2 m, respectively. This product can be expressed as the following matter-element:

$$\begin{bmatrix} \text{Wall lamp } D, & \text{height,} & 0.5\,\text{m} \\ & \text{width,} & 0.2\,\text{m} \end{bmatrix}.$$

If the following substituting transformation of the characteristics is applied,

$$T\begin{bmatrix} \text{Wall lamp } D, & \text{height,} & 0.5\,\text{m} \\ & \text{width,} & 0.2\,\text{m} \end{bmatrix} = \begin{bmatrix} \text{Wall lamp } D, & \text{width,} & 0.5\,\text{m} \\ & \text{height,} & 0.2\,\text{m} \end{bmatrix},$$

then an idea of a wall lamp with a width of 0.5 m and a height of 0.2 m can be obtained. The essence of implementing the transformation is to obtain a new wall lamp upon horizontally placing the original wall lamp.

Example 5.1.9

According to the divergence rules, the functional affair-element extended in Example 5.1.6 can also be extended to

$$
\left(\text{Control, dominating object, light}\right)\dashv
\begin{cases}
\left(\text{Provide, dominating object, light}\right) \\
\qquad \dashv \begin{cases}
\left(\text{Provide, dominating object, heat}\right) \\
\left(\text{Beautify, means, light}\right) \\
\left(\text{Lighten, tools, light}\right)
\end{cases} \\
\left(\text{Adjust, dominating object, light}\right) \\
\qquad \dashv \left(\text{Adjust, dominating object, temperature}\right) \\
\left(\text{Shield, dominating object, light}\right) \\
\qquad \dashv \left(\text{Shield, dominating object, mosquito}\right) \\
\left(\text{Strengthen, dominating object, light}\right) \\
\qquad \dashv \left(\text{Strengthen, dominating object, color}\right) \\
\left(\text{Filtrate, dominating object, light}\right) \\
\qquad \dashv \left(\text{Filtrate, dominating object, impurities}\right) \\
\left(\text{Absorb, dominating object, light}\right) \\
\qquad \dashv \left(\text{Absorb, dominating object, water}\right)
\end{cases}.
$$

This means that more substituting transformations of functional affair-elements can be applied, such as

$$T_1\left(\text{Control, dominating object, light}\right)=\left(\text{Beautify, means, light}\right),$$

$$T_2\left(\text{Control, dominating object, light}\right)=\left(\text{Strengthen, dominating object, color}\right).$$

QUESTIONS AND THINKING:

■ What would you think of upon observing Figures 5.1.4 and 5.1.5?
■ What kinds of transformations are applied from the left figure to the right one? Use a wall lamp D as an example to illustrate the transformations.

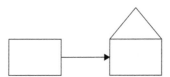

Figure 5.1.4   Schematic diagram of adding transformation.

Figure 5.1.5   Schematic diagram of removing transformation.

5.1.2 Adding/Removing Transformation of Basic-elements

5.1.2.1 Adding/removing Transformation of Values of Basic-elements

The value on a certain characteristic for an object of a certain basic-element is transformed into another value through the adding/removing transformation, denoted as:

$$T_1(O, \quad c, \quad v) = (O', \quad c, \quad v \oplus v_0),$$

$$T_2(O, \quad c, \quad v) = (O'', \quad c, \quad v \ominus v_0).$$

Illustration:

1. Adding/removing transformations are implemented on the value of a basic-element. It generally leads to conductive transformation of the object (for details, refer to Section 5.3). Detailed discussions on such transormations will be not conducted in this book. In the same manner, adding/removing transformations are also implemented on the object of a basic-element. That generally leads to conductive transformation of certain characteristics.
2. According to the combination rule in Section 3.4, the adding transformation of a basic-element can also be subdivided into sum adding transformation ($\oplus$) and integral adding transformation ($\otimes$). In this book, however, both are denoted as $\oplus$ for the sake of simplicity. If differentiation is necessary, there will be illustrations in relevant case studies. Similarly, the removing transformation is denoted with $\ominus$ in this book.

5.1.2.2 Adding/removing Transformation on the Objects and on the Values of Basic-elements

The object of a certain basic-element and its value about a certain characteristic is transformed into another object and value by the adding/removing transformation, denoted as:

$$T_1(O, \quad c, \quad v) = (O \oplus O_0, \quad c, \quad v \oplus v_0) \quad \text{or}$$

$$T_2(O, \quad c, \quad v) = (O \ominus O_0, \quad c, \quad v \ominus v_0).$$

Case Analysis:

Example 5.1.10

If a cuboid wall lamp D is on the left side of Figures 5.1.4 and 5.1.5, then on the right side of Figure 5.1.4 is a lamp adding a cone D', while on the right side of Figure 5.1.5 is a lamp deleting a cone D''. These statements can be formalized as follows:

$$T_1\big(D, \quad \text{shape}, \quad \text{cuboid}\big) = \big(D, \quad \text{shape}, \quad \text{cuboid} \oplus \text{ellipsoid}\big)$$

$$T_2\big(D, \quad \text{shape}, \quad \text{cuboid}\big) = \big(D'', \quad \text{shape}, \quad \text{cuboid} \ominus \text{ellipsoid}\big).$$

Example 5.1.11

The wall lamp has the function of lighting, while the bouquet and the stereo have functions of aesthetics and playing music, respectively. Thereby, a creative idea with combined functions can be obtained for a new product by the adding transformation.

A product's functions can be expressed in the formalization of affair-elements, while the essences of "illuminate" and "aesthetics" are to "provide light" and to "beautify the environment." Therefore, the three functions in this example can be expressed by using affair-elements as follows:

$$A = \begin{bmatrix} \text{Provide,} & \text{dominating boject,} & \text{light} \\ & \text{means,} & \text{wall lamp } D \end{bmatrix}$$

$$A_1 = \begin{bmatrix} \text{Beautify,} & \text{dominating boject,} & \text{environment} \\ & \text{means,} & \text{bouquet } D_1 \end{bmatrix}$$

$$A_2 = \begin{bmatrix} \text{Play,} & \text{dominating boject,} & \text{music} \\ & \text{means,} & \text{stereo } D_2 \end{bmatrix}.$$

For this reason, the following adding transformations can be implemented to obtain combined functions, and a new creative idea for the product:

$$T_1A = A \oplus A_1 = \begin{bmatrix} \text{Provide} \oplus \text{Beautify,} & \text{dominating boject,} & \text{light} \oplus \text{environment} \\ & \text{means,} & \text{wall lamp } D \oplus \text{bouquet } D_1 \end{bmatrix}$$

$$T_2A = A \oplus A_2 = \begin{bmatrix} \text{Provide} \oplus \text{Play,} & \text{dominating boject,} & \text{light} \oplus \text{music} \\ & \text{means,} & \text{wall lamp } D \oplus \text{stereo } D_2 \end{bmatrix}.$$

Example 5.1.12

A certain crystal lamp D_1 dedicated for the living room has a basic function of "providing light", which can be expressed as a functional affair-matter:

$$\begin{bmatrix} \text{Provide,} & \text{dominating object,} & \text{light} \\ & \text{means,} & D_1 \end{bmatrix}.$$

The LED lamp bead D_2 with a function of "adding colors" can be implemented as an adding transformation of the functional affair-elements:

$$T_1 \begin{bmatrix} \text{Provide,} & \text{dominating object,} & \text{light} \\ & \text{means,} & D_1 \end{bmatrix}$$

$$= \begin{bmatrix} \text{Provide,} & \text{dominating object,} & \text{light} \\ & \text{means,} & D_1 \end{bmatrix} \oplus \begin{bmatrix} \text{Add,} & \text{dominating object,} & \text{colors} \\ & \text{means,} & D_2 \end{bmatrix}$$

$$= \begin{bmatrix} \text{Provide} \oplus \text{Add,} & \text{dominating object,} & \text{light} \oplus \text{colors} \\ & \text{means,} & D_1' \end{bmatrix}.$$

A certain wall lamp has the basic function of "providing light for people D_3", while people also have the need of "charging mobile phone D_4." As people's need is corresponding to the product function, the wall lamp can be implemented as an adding transformation of the functional affair-elements as follows:

$$T_2 \begin{bmatrix} \text{Provide,} & \text{dominating object,} & \text{light} \\ & \text{means,} & D_2 \\ & \text{receiving object,} & D_3 \end{bmatrix}$$

$$= \begin{bmatrix} \text{Provide,} & \text{dominating object,} & \text{light} \\ & \text{means,} & D_2 \\ & \text{receiving object,} & D_3 \end{bmatrix} \oplus \begin{bmatrix} \text{Supplement,} & \text{dominating object,} & \text{power} \\ & \text{receiving object,} & D_4 \end{bmatrix}$$

$$= \begin{bmatrix} \text{Provide} \oplus \text{Supplement,} & \text{dominating object,} & \text{light} \oplus \text{power} \\ & \text{means,} & D_2' \\ & \text{receiving object,} & D_3 \oplus D_4 \end{bmatrix}.$$

Namely, a wall lamp with a charging function (e.g., with a USB charging interface) can be designed.

Illustration: In particular, the adding transformation that is conducted among any two basic-elements without a priority is called combination transformation, such as three-bulb dining room lamp is a mix of light bulbs without priority; a two-bulb wall lamp, a two-bulb lamp, a multi-bulb lamp, combination furniture, combination stationery, and so on, are cases of combination transformations.

Example 5.1.13

In certain fast food restaurants, Coca-Cola D_1, burger D_2 and fries D_3 are often sold as a combination meal, which is often cheaper than buying all three separately. The cost after combination is equal to the sum of costs of all components. Such a combination transformation can be expressed as an extension transformation:

$$
T_1 \left\{
\begin{bmatrix} D_1, & \text{cost,} & v_{11} \\ & \text{price,} & v_{12} \end{bmatrix}
\oplus
\begin{bmatrix} D_2, & \text{cost,} & v_{21} \\ & \text{price,} & v_{22} \end{bmatrix}
\oplus
\begin{bmatrix} D_3, & \text{cost,} & v_{31} \\ & \text{price,} & v_{32} \end{bmatrix}
\right\}
$$

$$
= \begin{bmatrix} D_1 \oplus D_2 \oplus D_3, & \text{cost,} & v_1 \\ & \text{price,} & v_2 \end{bmatrix}
$$

where $\sum_{i=1}^{3} v_{i1} = v_1, \sum_{i=1}^{3} v_{i2} > v_2$.

If adding small toys D_4 to the combination meal package and the total price is not changed, it can better attract children to buy.

$$
T_2 \left\{
\begin{bmatrix} D_1, & \text{cost,} & v_{11} \\ & \text{price,} & v_{12} \end{bmatrix}
\oplus
\begin{bmatrix} D_2, & \text{cost,} & v_{21} \\ & \text{price,} & v_{22} \end{bmatrix}
\oplus
\begin{bmatrix} D_3, & \text{cost,} & v_{31} \\ & \text{price,} & v_{32} \end{bmatrix}
\oplus
\begin{bmatrix} D_4, & \text{cost,} & v_{41} \\ & \text{price,} & v_{42} \end{bmatrix}
\right\}
$$

$$
= \begin{bmatrix} D_1 \oplus D_2 \oplus D_3 \oplus D_4, & \text{cost,} & v_1' \\ & \text{price,} & v_2 \end{bmatrix}
$$

where $\sum_{i=1}^{4} v_{i1} = v_1', \sum_{i=1}^{4} v_{i2} > v_2$.

Questions and Thinking:

- What would you think upon observing Figures 5.1.6 and 5.1.7 next page?
- What kinds of transformations are applied from the figures on the left to the figures on the right? Use wall lamp D as an example to illustrate the transformations.

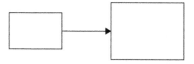

Figure 5.1.6   Schematic diagram of enlarging transformation.

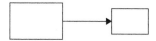

Figure 5.1.7   Schematic diagram of shrinking transformation.

5.1.3 Enlarging/Shrinking Transformation of Basic-elements

5.1.3.1 Enlarging/shrinking Transformation on Values of Basic-elements

The value of a characteristic for an object of a certain basic-element is transformed into another value through the enlarging or shrinking transformation, denoted as:

$$T_1\big(O,\ c,\ v\big)=\big(O',\ c,\ \alpha v\big),\quad \alpha>1,$$

$$T_2\big(O,\ c,\ v\big)=\big(O'',\ c,\ \alpha v\big),\quad 0<\alpha<1.$$

Illustration: The enlarging/shrinking transformation of the value will inevitably lead to the enlarging/shrinking of the object.

5.1.3.2 Enlarging/shrinking Transformation on Objects of Basic-elements

The object of a basic-element is transformed into another object and value through the enlarging/shrinking transformation, denoted as:

$$T_1\big(O,\ c,\ v\big)=\big(\alpha O,\ c,\ v'\big),\quad \alpha>1,$$

$$T_2\big(O,\ c,\ v\big)=\big(\alpha O,\ c,\ v''\big),\quad 0<\alpha<1.$$

Explanation: An enlarging/shrinking transformation of an object must definitely be an enlarging/shrinking transformation occurring in the value of a certain characteristic, but it can not definitely result in an enlargement/shrinkage of the value on all the characteristics of the object.

Case Analysis:

Example 5.1.14

An enlarging/shrinking of the value of the illuminating angle of the wall lamp D_1 may lead to different effects of light patterns; an enlarging/shrinking of the value of the length of the ceiling lamp D_2 is suitable for rooms of different areas. Taking an enlarging transformation as an example, the following transformations can be expressed as:

$$T_1\left[D_1, \quad \text{illuminating angle}, \quad 45° \right] = \left[D_1', \quad \text{illuminating angle}, \quad 135° \right],$$

$$T_2\left[D_2, \quad \text{length}, \quad 50 \text{ cm} \right] = \left[D_2', \quad \text{length}, \quad 100 \text{ cm} \right].$$

Example 5.1.15

Children's candy is commonly packaged in plastic bags, tin boxes, paper boxes, etc. Is it possible to obtain new creative ideas of packaging by using an enlarging and duplicating transformation?

$$T \begin{bmatrix} \text{Packaging paper } D, & \text{width}, & v_1 \\ & \text{length}, & v_2 \\ & \text{material}, & \text{plastic} \\ & \text{pattern}, & \text{cat} \\ & \text{shape}, & \text{rectangle} \end{bmatrix}$$

$$= \begin{bmatrix} \text{Packaging paper } D', & \text{width}, & n \cdot v_1 \\ & \text{length}, & n \cdot v_2 \\ & \text{material}, & \text{plastic} \\ & \text{pattern}, & \text{cat} \\ & \text{shape}, & \text{rectangle} \end{bmatrix}.$$

Creative ideas can be obtained from this transformation: a wrapper for a sweet can be expanded into a larger wrapper for n sweets, while the original material, pattern, and shape are retained.

Figure 5.1.8 Case diagram of enlarging transformation.

A cylindrical carton support should be added for supporting the inside of the plastic paper to make the shape of packaging transformed slightly from the original one. It is a factor that should be considered when creative ideas are formed into a program (Figure 5.1.8).

QUESTIONS AND THINKING:

- What would you think about upon observing Figure 5.1.9?
- What kind of transformation is employed from the left figure to the right one?

5.1.4 Decomposing Transformation of Basic-elements

A decomposing transformation of a matter-element decomposes a value of a certain characteristic of an object into multiple values. Correspondingly, the object is also divided into multiple objects.

There are two kinds of decomposing transformations for affair-elements.

1. A value about a certain characteristic of a certain affair-element can be decomposed into multiple values; correspondingly, an action is also divided into multiple actions.
2. An action of a certain affair-element can be decomposed into multiple actions; accordingly, a value about a certain characteristic is decomposed into multiple values.

In either situation, there are exceptional cases in which the adopted decomposing transformation is not performed on a corresponding action or value. A decomposing transformation of relation-elements is similar to that of affair-elements, which will not be discussed in detail.

In general, the general decomposing transformation can be expressed as

$$T\left(O, \quad c, \quad v\right) = \left\{\left(O_1, \quad c, \quad v_1\right), \left(O_2, \quad c, \quad v_2\right), \cdots, \left(O_n, \quad c, \quad v_n\right)\right\}.$$

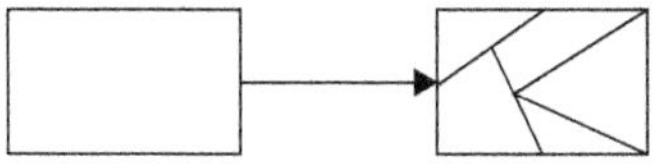

Figure 5.1.9 Schematic diagram of decomposing transformation.

In Figure 5.1.9, the figure on the right is obtained by employing a decomposing transformation of the figure on the left side. Regarding different characteristics, different composition results can be obtained. For example, the decomposing transformation on a shape's value conducted for an object O can be expressed as

$$T_1(O, \text{ shape, rectangle}) = \{(O_1, \text{ shape, right triangle}), (O_2, \text{ shape, trapezoid}),$$

$$(O_3, \text{ shape, trapezoid}), (O_4, \text{ shape, equilateral triangle}),$$

$$(O_5, \text{ shape, obtuse triangle})\}.$$

The decomposing transformation on a color's value can be expressed as

$$T_2(O, \text{ color, white}) = \{(O_1, \text{ color, white}), (O_2, \text{ color, white}),$$

$$(O_3, \text{ color, white}), (O_4, \text{ color, white}), (O_5, \text{ color, white})\}.$$

If the substituting transformation on a color's value is conducted on each matter-element of the above compositions (a transformation conducted repeatedly will be illustrated in the operation of transformations), then the following results can be obtained:

$$T_3 T_2(O, \text{ color, white}) = \{(O_1, \text{ color, red}), (O_2, \text{ color, yellow}),$$

$$(O_3, \text{ color, blue}), (O_4, \text{ color, purple}), (O_5, \text{ color, gray})\}.$$

A decomposing transformation on a weight's value can be expressed as

$$T_4(O, \text{ weight, 500 g}) = \{(O_1, \text{ weight, 100 g}), (O_2, \text{ weight, 80 g}),$$

$$(O_3, \text{ weight, 120 g}), (O_4, \text{ weight, 150 g}), (O_5, \text{ weight, 50 g})\}.$$

Note: The sum of values of different characteristics after decomposing is not definitely equal to the original value. For example, if a used automobile is sold after dismantling, the sum of selling prices of parts will be larger than that of the overall sale of the automobile as a whole. Thereby, income can be increased. Namely, we have

$$T_5(\text{Car } D, \text{ selling price, } v)$$

$$= \{(\text{Tire } D_1, \text{ selling price, } v_1), (\text{Engine } D_2, \text{ selling price, } v_2),$$

$$(\text{Alternator } D_3, \text{ selling price, } v_3), (\text{Power supply } D_4, \text{ selling price, } v_4), \cdots\}.$$

Figure 5.1.10　Tangram before transformations.

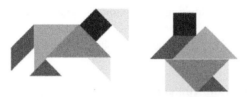

Figure 5.1.11　Tangram after transformations.

EXTENDED THINKING:

- What would you think of upon observing Figures 5.1.10 and 5.1.11?
- What kind of transformation happens from the left figure to the right one?

There are a variety of situations for possibly employing decomposing transformations, and different situations tend to require different decomposing modes.

Decomposing transformation about time parameters: If t stands for the time parameter, then we have

$$T\big(O(t),\ c,\ v(t)\big) = \big\{\big(O(t_1),\ c,\ v(t_1)\big),\big(O(t_2),\ c,\ v(t_2)\big),\ldots,\big(O(t_n),\ c,\ v(t_n)\big)\big\}.$$

For example, the light color of the original ceiling lamp $O(t)$ is yellow. Light of different colors can be emitted by the ceiling lamp $O(t)$ at different times by the following decomposing transformation. Suppose $c =$ color of light, then

$$T_1\big(O(t),\ c,\ \text{yellow}\big)$$

$$= \big\{\big(O(t_1),\ c,\ \text{yellow}\big),\big(O(t_2),\ c,\ \text{green}\big)\cdots,\big(O(t_n),\ c,\ \text{red}\big)\big\}.$$

Decomposing transformation about other parameters: If t represents a parameter other than time, such as scenarios and locations, then we have

$$T\big(O(t),\ c,\ v(t)\big) = \big\{\big(O(t_1),\ c,\ v(t_1)\big),\big(O(t_2),\ c,\ v(t_2)\big),\ldots,\big(O(t_n),\ c,\ v(t_n)\big)\big\}.$$

For example, the light intensity of the original ceiling lamp $O(t)$ is constant. Yet the light intensity can be adjusted according to different scenarios, such as reading, resting, and looking at the phone, by a decomposing transformation, which can be formalized as follows:

$$T_2\Big(\text{Ceiling lamp } O(t), \quad \text{light intensity}, \quad \text{bright}\Big)$$

$$= \Big\{\big(\text{Ceiling lamp } O(\text{reading}), \quad \text{light intensity}, \quad \text{bright}\big),$$

$$\big(\text{Ceiling lamp } O(\text{looking at the phone}), \quad \text{light intensity}, \quad \text{medium}\big),$$

$$\big(\text{Ceiling lamp } O(\text{rest}), \quad \text{light intensity}, \quad \text{weak}\big)\Big\}$$

QUESTIONS AND THINKING:

- What would you think about upon observing Figure 5.1.12?
- What kind of transformation is happening from the left figure to the right one?

5.1.5 Duplicating Transformation of Basic-elements

A duplicating transformation is a special kind of basic transformation, such as photo-printing, copying, scanning, printing, burning of discs, recording, video recording, methods for repeated uses and duplicating of products, etc. Such transformations are widely used in the information field.

Mass production is also a kind of duplication. Duplication includes both material and non-material duplication. Conditions provided can be divided into two classes: ones that can be used repeatedly, and ones that cannot be duplicated yet can only be used through distribution.

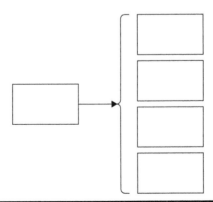

Figure 5.1.12 Schematic diagram of duplicating transformation.

A duplicating transformation is denoted as $TB = \{B, B^*\}$.

Duplicating transformations can be divided into various types. The object becomes at least two objects—i.e. the original object and a duplicated object—after the implementation of a duplicating transformation. It can also become multiple objects. Depending on what objects are duplicated, duplicating transformations are divided into the following categories:

Enlarging duplicating transformation: $TB = \{B, \alpha B\}, \alpha > 1$;

Shrinking duplicating transformation: $TB = \{B, \alpha B\}, 0 < \alpha < 1$;

Approximate duplicating transformation: $TB = \{B, B^*\}, B \approx B^*$, where "$\approx$" is an approximation; and

Multiple duplicating transformation: $TB = \{B, B_1^*, B_2^*, \cdots, B_n^*\}$.

Illustration: According to the conjugate analysis in Chapter 4, those duplicating transformations that change non-material objects to material objects, or material objects to non-material objects, are classified as conjugate transformations, which will be discussed in Section 5.4. For example, a three-dimensional printer is produce the figure of E as the material object E', such that one of the two figures is non-material while the other is material. That is, we have

$$T \begin{bmatrix} \text{Figure } E_{im}, & \text{measurement,} & v_1 \\ & \text{color,} & v_2 \end{bmatrix}$$

$$= \left\{ \begin{bmatrix} \text{Figure } E_{im}, & \text{measurement,} & v_1 \\ & \text{color,} & v_2 \end{bmatrix}, \begin{bmatrix} \text{Material object } E_{re}, & \text{measurement,} & v_1 \\ & \text{color,} & v_2' \end{bmatrix} \right\}.$$

Summary

The basic transformation methods for basic-elements include substituting, adding/removing, enlarging/shrinking, decomposing, duplicating, and other transformations. Because each basic-element is a triad that is composed of objects, characteristics, and values, these transformations can be divided into transformations of various elements in the triad. The source of the transformation is based on the extensible analysis.

1. For a substituting transformation, the basic-element that can substitute B may be found from the divergent basic-element set of B.
2. For an adding transformation, the basic-element combining with B can be found from a combined basic-element set of B.

3. For a removing transformation, the separable basic-element can only be transformed through removing after analyzing whether B is a separable basic-element.
4. For an enlarging transformation, the object or value of the basic-element B must be enlargable.
5. For a shrinking transformation, the object or value of the basic-element B must be shrinkable.
6. For a decomposing transformation, the basic-element B can be decomposed into multiple basic-elements after analyzing and confirming that the basic-element B is a separable basic-element.

Moreover, because there are inextricable connections among the three elements of a basic-element, applying a transformation on some elements may lead to corresponding transformations of other elements. This is called a conductive transformation (see Section 5.3).

QUESTIONS AND THINKING:
- Is it an innovation to change rules? Can contradictory problems be solved by changing the given rules?
- Is it possible to change the functional relation of relevant basic-elements?
- Is it possible to compulsorily establish a brand new relevant?

5.1.6 Basic Transformation Methods of Rules

Rules, also known as criteria, are important conditions for innovating or solving contradictory problems. For example, the required diameter range for producing a ceiling lamp of 50 cm in diameter is <49.9, 50.1> cm. As another example, employers may require future employees to have a bachelor degree or more advanced degrees in computer-relevant majors and have obtained a certain level of professional certification for people applying for information technology positions. These are imposed rules.

In practice, these rules are changeable. Because problems can become contradictory due to the introduction of inappropriate rules, rule changing can contribute to the success of solving these contradictory problems. As for product innovation, new product ideas may be generated by changing rules.

The general representation of the basic transformation of a rule is $T_k k = k'$. Basic transformation methods of rules include:

1. The substituting transformation method utilizes a new rule to replace the original rule: $T_k k = k'$;
2. The adding transformation method adds a new rule to the set of the original rules: $T_k k = k \oplus k_1$;

3. The removing transformation method deletes or decreases a part of the requirement of the original rule: $T_k k = k \ominus k_1$;
4. The number-enlarging transformation method expands the original rule by the amount of multiples: $T_k k = \alpha k, \alpha > 1$;
5. The number-shrinking transformation method decreases the original rule by the amount of multiples: $T_k k = \alpha k, 0 < \alpha < 1$; and
6. The decomposing transformation method divides the original rule into more detailed rules to allow different rules to be applied to different objects: $T_k k = \{k_1, k_2, \cdots, k_n\}$.

The study of transformations of rules involves transforming the mapping relationship between elements and real numbers, ushering in a new path for innovation or resolution of contradictory problems.

Case Analysis:

Example 5.1.16

The value of the "tilt angle" of a wall lamp is a function of that of the value of the "irradiation angle", or it can be said that the "tilt angle" determines the "irradiation angle." Is it possible to change this rule?

Assume that $M_1 =$ (Wall lamp D, inclination angle, v_1), $M_2 =$ (Wall lamp D, inclination angle, v_2). According to the particular domain knowledge, $v_2 = k(v_1)$ can be obtained. According to relevant rules in Chapter 4, we have $M_1 \sim M_2$.

If the substituting transformation $T_k k = k'$ is conducted or the removing transformation $T_k k = k - k = 0$ is implemented on rules, then the functional relationship between the values of the structure and the irradiation angle of the wall lamp is transformed into another relation or irrelevance. In such cases, the transformed wall lamp is a brand new wall lamp.

These two situations—before and after transformations—can be expressed as matter-elements.

$$M = \begin{bmatrix} \text{Wall lamp D,} & \text{inclination angle,} & v_1 \\ & \text{irradiation angle,} & k(v_1) \end{bmatrix}$$

$$M' = \begin{bmatrix} \text{Wall lamp D',} & \text{inclination angle,} & v_1 \\ & \text{irradiation angle,} & k'(v_1) \end{bmatrix}$$

$$M'' = \begin{bmatrix} \text{Wall lamp D'',} & \text{inclination angle,} & v_1 \\ & \text{irradiation angle,} & v_2'' \end{bmatrix}$$

Example 5.1.17

The value of "brightness" or "color" of a bedside lamp has no correlation to that of the "position" of bed and the user's "mood." Is it possible to establish a correlation?

Set M_{11} = (Bedside lamp D_1, brightness, v_{11}), M_{12} = (Bedside lamp D_1, color, v_{12}),

M_2 = (Bed D_2, position, v_2), and M_3 = (User D_3, mood, v_3).

The statement that "light changes as one wishes" means that the "color" and the "brightness" of the light change with people's mood. For such a situation to happen, it requires the establishment of a correlation between the values of the two characteristics, i.e. to establish a new rule:

$$T_{11}v_{11} = v_{11}' = k_1(v_2) \wedge k_2(v_3), \quad T_{12}v_{12} = v_{12}' = k_3(v_3).$$

The following matter-element can be used to express the situation implemented in this new rule:

$$M_1' = \begin{bmatrix} M_{11}' \\ M_{12}' \end{bmatrix} = \begin{bmatrix} D_1', & \text{brightness}, & v_{11}' \\ & \text{color}, & v_{12}' \end{bmatrix}$$

$$= \begin{bmatrix} D_1', & \text{brightness}, & k_1(v_2) \wedge k_2(v_3) \\ & \text{color}, & k_3(v_3) \end{bmatrix}.$$

EXTENDED THINKING:

■ Is it possible to establish an affiliation between light fixture and smart appliances?
■ How can more correlations be established?

Example 5.1.18

There is no correlation between the value of "brightness" or "color" of the bedside lamp and the "use time." Is it possible to establish a correlation?

There is no correlation between the value of "tilt" of the bedside lamp and that of the user's "use state." Is it possible to establish a correlation?

If functional relations k_1, k_2, k_3 are established among the values of these characteristics, the following matter-element can be used for expressing the situations before and after the transformation:

$$\text{Set } M_1 = \begin{bmatrix} \text{Bedside lamp } D_1, & \text{brightness}, & v_{11} \\ & \text{color}, & v_{12} \\ & \text{inclination angle}, & v_{13} \end{bmatrix}, M_2 = \begin{bmatrix} \text{User } D_2, & \text{use time}, & v_{21} \\ & \text{use state}, & v_{22} \end{bmatrix}.$$

Implement the transformation:

$$T_1 v_{11} = v'_{11} = k_1(v_{21}), \quad T_2 v_{12} = v'_{12} = k_2(v_{21}), \quad T_3 v_{13} = v'_{13} = k_3(v_{22}).$$

The matter-element of the bedside lamp after transformation is

$$M = \begin{bmatrix} \text{Bedside lamp } D'_1, & \text{brightness,} & v'_{11} \\ & \text{color,} & v'_{12} \\ & \text{inclination angle,} & v'_{13} \end{bmatrix} = \begin{bmatrix} \text{Bedside lamp } D'_1, & \text{brightness,} & k_1(v_{21}) \\ & \text{color,} & k_2(v_{21}) \\ & \text{inclination angle,} & k_3(v_{22}) \end{bmatrix}.$$

QUESTIONS AND THINKING:

- Is it possible to change the field of application while either innovating or resolving contradictions?
- Can products applied in the military field be transformed for civilian applications?
- Can products applied in a family home setting be transformed for use in public places or outdoors?

5.1.7 Basic Transformation Methods of Domain (Field)

Basic transformations of one domain include substituting transformation, adding transformation, removing transformation, and decomposing transformation. When dealing with the real number field, the enlarging or shrinking transformations can be performed on the domain. The domain of discourse is considered to be fixed and unchanged in the classic set and fuzzy set theories, while we consider that the domain can be transformed in extension sets, providing a new path to resolve contradictory problems (for details, see Chapter 7).

For example, large multinationals tend to move their production bases from developed countries to developing countries with lower labor and resource costs in the global wave of economic integration, or they realize a strategic transformation from "production and selling place" to "production in the selling place," which is a substituting transformation of the domain of discourse. In terms of expanding the scope of use objects for a certain product, expanding the original single object O to O_1, O_2, O_3, O_4, is an enlarging transformation of the domain of discourse. If the original use object of a certain product is a baby, expanding to children, women, etc., is the enlarging transformation of the domain of discourse. The change of the sales scope of a certain product from the original area A to the province, the country, and even abroad, is also an enlarging transformation of the domain of discourse.

The greatest enlightenment from domain transformations is that we shouldn't "consider the matter as it stands" in the process of dealing with contradictory problems or innovative thinking. Instead, we should try our best to conduct substituting transformations, enlarging transformations, or shrinking transformations on the objects to be observed so as to break through the contradictions of the original problems. In this way, a creative idea will be obtained.

According to the extensible analysis of a domain, basic transformations of the domain are as follows.

(1) **Substituting transformation:** For any given domain U, if another domain U' and the transformation T exist such that $T_U U = U'$, then the transformation T is a substituting transformation of the domain U. For example, if an enterprise's target market is city A, then its domain U is all of the people in city A. When the target market is transformed to another place, it is equivalent to a substituting transformation of the domain.

(2) **Adding transformation:** For any given domain U, if another domain U_1 and a transformation T exist such that $T_U U = U \cup U_1$, then the transformation T is called an adding transformation of the domain U.

(3) **Removing transformation:** For any given domain U, if another domain U_1 and a transformation T exist such that $T_U U = U_1$, $U_1 \subset U$, then the transformation T is called a removing transformation of the domain U. For example, a certain enterprise's market domain U is the set of all of the people in city A. When this domain cannot meet the needs of the enterprise, an adding transformation of the domain can be adopted. For example, $U_1 = \{$all people in city B$\}$ is taken as $TU = U \cup U_1$, representing an adding transformation of the domain. When the enterprise adjusts the product structure to produce only products used by some special groups, a removing transformation of the domain can be adopted. For example, a certain garment enterprise produces only clothing for students. The domain of the enterprise consists of students only, i.e. $U_1 = \{$all students in city A$\}$, $U_1 \subset U$, rather than considering to the domain U of all people in city A.

(4) **Number-enlarging transformation:** For a real number domain U, if a transformation T and a real number $\alpha(\alpha > 1)$ exist such that $T_U U = \alpha U$, then the transformation T is called a number-enlarging transformation of the real number domain U.

(5) **Number-shrinking transformation:** For a real number domain U, if a transformation T and a real number $\alpha(0 < \alpha < 1)$ exist such that $T_U U = \alpha U$, then the transformation T is called a number-shrinking transformation of the real number domain U.

(6) **Decomposing transformation:** For any domain U, if a transformation T exists such that $T_U U = \{U_1, U_2, ..., U_n\}$, satisfying $U_i \subset U$ $(i = 1, 2, ..., n)$, then the transformation T is called a decomposing transformation of the domain U.

Case Analysis:

Example 5.1.19

With the development of society, more and more robotic arms have entered everyday life. Robotic arms made by a certain automation company have been sold mainly to local developed industrialized enterprises. The automation company seeks to create market-exploring ideas by utilizing transformations of domains.

We must first determine the original domain of the problem under consideration. The domain of the problem to be studied is $U = \{$Enterprises in developed industrial areas$\}$. According to the enterprise's situation, the company's sales is based only on the geographical advantage.

Once the domain is determined, we must select transformations for the domain.

1. **For the substituting transformation of the domain:** Because new products tend to only tap a small portion of the market in its original domain of discourse, let us implement $T_1U = U_1$. That is, the original domain is abandoned, and an economically developed industrial city that is similar to the local industry development is selected to be a new domain U_1. Markets are explored on the basis of the new domain. Sales in domain U_1 is a consequence of the thinking logic of "sending out for selling."

2. **For the adding/removing transformation of the domain:** $T_2U = U \cup U' = U_2$. In other words, neighboring provinces are also considered as part of the domain on the basis of the original domain. If a removing transformation of the domain is implemented on U_2, $T_3U_2 = U_3$, $(U_3 \subset U_2)$, that is, if U_3 is a subset of U_2, such as local agricultural users, good sales performance can be created by advertising the robot arm's characteristics of combining automation and semi-automation.

3. **For the decomposing transformation of the domain:** Different mechanical arms will be produced and different sales modes will be implemented to target different agricultural plants. A transformation can be used on the original domain, or in a substituted domain, or in an expanded domain such as $T_4U = \{U'_1, U'_2, U'_3\}$, where $U'_1 = \{$all large industrial companies in $U\}$, $U'_2 = \{$all small and medium-sized industrial companies in $U\}$, and $U'_3 = \{$all farmers in $U\}$. The automation company can comprehensively adopt appropriate transformations according to such factors as adjusted promotion costs, shipping costs, sales volume forecasts, price/cost, etc.

Example 5.1.20

The application domain of the general bedside lamp is for civil use, and the users are residents, i.e. the domain $U = \{$all residents$\}$; the place of use is households of families, and the way of use is to place it on the nightstand by the bed. If a substituting transformation is performed on the area of application by changing its civil use to military use, then the substituting transformation $TU = U' = \{$soldiers$\}$ can be made. Accordingly, values about other characteristics should be accordingly transformed (i.e. a conductive transformation that will be discussed in Section 5.4). Specific methods of transformation should be analyzed case by case. The situations before and after the transformation can be expressed in the following matter-elements.

$$M' = \begin{bmatrix} \text{Bedside lamp } D, & \text{brightness,} & v_1' \\ & \text{color,} & v_2' \\ & \text{application domain,} & \text{civil use} \\ & \text{user,} & \text{resident} \\ & \text{using place,} & \text{family} \\ & \text{using way,} & \text{bedside table} \end{bmatrix} \text{ and}$$

$$M' = \begin{bmatrix} \text{Bedside lamp } D, & \text{brightness,} & v_1' \\ & \text{color,} & v_2' \\ & \text{application domain,} & \text{military use} \\ & \text{user,} & \text{soldiers} \\ & \text{using place,} & \text{field} \\ & \text{using way,} & \text{fastened to the tent} \end{bmatrix}.$$

5.2 Operation Methods of Extension Transformations

QUESTIONS AND THINKING:

- Would you like to use basic transformations mentioned in the previous section to compose other complex transformations to help you solve problems orderly?
- Have you ever encountered any example requiring continuous implementation of multiple transformations to solve problems?
- Have you ever encountered any example requiring simultaneous implementation of multiple transformations to solve problems?
- What other transformations have you encountered that differ from the basic examples introduced in the previous section?

In this section, commonly used operations of transformations, including PRODUCT transformation, AND transformation, OR transformation, and INVERSE transformation will be discussed.

5.2.1 PRODUCT Transformation Method

For a certain object (respectively, basic-element, rule, or domain of discourse) B_0, if transformation T_1 and transformation T_2 exist on the basis of $T_1 B_0 = B_1$, $T_2 B_1 = B_2$, then $T B_0 = T_2 (T_1 B_0) = T_2 B_1 = B_2$ and the transformation $T = T_2 T_1$ is known as the PRODUCT of transformations T_1 and T_2. The PRODUCT transformation involves the continuous implementation of two or more transformations on a certain object.

Case Analysis:

Example 5.2.1

To move an object D from the first floor to the thirtieth floor, you must first move it into the elevator. With the elevator you move D from the first floor to the thirtieth floor. Then you move D into the predetermined room from the elevator. That is, we have

$$T_1\big(D, \quad \text{position}, \quad \text{the first floor}\big) = \big(D, \quad \text{position}, \quad \text{elevator on the first floor}\big)$$

$$T_2\big(D, \quad \text{position}, \quad \text{elevator on the first floor}\big) = \big(D, \quad \text{position}, \quad \text{elevator on the 30th floor}\big)$$

$$T_3\big(D, \quad \text{position}, \quad \text{elevator on the 30th floor}\big) = \big(D, \quad \text{position}, \quad \text{room on the 30th floor}\big).$$

So, $T = T_3 T_2 T_1$ is the PRODUCT transformation of three transformations T_1, T_2, and T_3.

When solving problems, PRODUCT transformation is commonly used for turning B_0 into B_2. When turning B_0 into B_2 cannot be realized directly, find transformations T_1 and T_2 so that T_1 can change B_0 into B_1, and then T_2 can turn B_1 into B_2.

When applying the PRODUCT transformation, it is important to note that there is an ordered sequence of transformations between T_1 and T_2. The PRODUCT transformation method is usually applied in cases where conflicts need to be resolved using more than one method.

Example 5.2.2

The assembly line a plant is to transfer component D_1 from position a_1 to position a_2 of component D_2 for assembling. After that, the resultant product is transferred to position a_3 of component D_3 for assembling... Such flow of partially assembled products continues until the completion of the whole product D. What is utilized in this flow of partially assembled products is the PRODUCT transformation method. That is, suppose that

$$M_1 = \big(D_1, \quad \text{position}, \quad a_1\big), \cdots\cdots, M_n = \big(D_n, \quad \text{position}, \quad a_n\big).$$

Next use the following transformations:

$$T_1 M_1 = M_1 \oplus M_2 = \big(D_1 \oplus D_2, \quad \text{position}, \quad a_2\big) = M_2',$$

$$T_2 M_2' = M_2' \oplus M_3 = M_3',$$

$$\cdots\cdots$$

$$T_{n-1} M_{n-1}' = M_{n-1}' \oplus M_n = M_n' = \big(D, \quad \text{position}, \quad a_n\big).$$

Finally, a product D, made up of n components, at position a_n is completely assembled by the transformation $T = T_{n-1} T_{n-2} \cdots T_2 T_1$.

5.2.2 AND Transformation Method

If transformations T_1 and T_2 exist simultaneously such that $T_1 B_1 = B_1'$, $T_2 B_2 = B_2'$ with $B_1' \wedge B_2' = B'$ (i.e. we have $T_1 B_1 \wedge T_2 B_2 = B_1' \wedge B_2' = B'$), then the transformation $T = T_1 \wedge T_2$ is known as the AND transformation of the transformations T_1 and T_2. The AND transformation involves the simultaneous implementation of two or more transformations on a certain object.

Case Analysis:

Example 5.2.3

In the design of a ceiling lamp, one can change the original glass material of the shell of the ceiling light D_1 into a plastic material and the original square shape into a circular shape obtain a creative idea of new products. These two transformations, when implemented at the same time, are an example of the AND transformation that can be expressed in the following model:

$$T_1 \left(D_1, \quad \text{material}, \quad \text{glass} \right) = \left(D_1', \quad \text{material}, \quad \text{plastic} \right),$$

$$T_2 \left(D_1, \quad \text{shape}, \quad \text{square} \right) = \left(D_1'', \quad \text{shape}, \quad \text{circular} \right).$$

So, the AND transformation is $T = T_1 \wedge T_2$, i.e.

$$T_1 \left(D_1, \quad \text{material}, \quad \text{glass} \right) \wedge T_2 \left(D_1, \quad \text{shape}, \quad \text{square} \right)$$

$$= \left(\begin{array}{ccc} D, & \text{material}, & \text{plastic} \\ & \text{shape}, & \text{circular} \end{array} \right).$$

Example 5.2.4

Suppose we have two pieces of plywood, each 10 cm thick with holes 10 cm in diameter, and they need to be connected with bolt D_2, but the specifications of bolt D_2 do not meet the required diameter and length. Therefore, the following transformations must be implemented when selecting a bolt:

$$T_1 \left(D_2, \quad \text{diameter}, \quad v_{21} \right) = \left(D_2', \quad \text{diameter}, \quad 10 \text{ cm} \right) \quad \text{and}$$

$$T_2 \left(D_2, \quad \text{length}, \quad v_{22} \right) = \left(D_2'', \quad \text{length}, \quad <15, 20> \text{cm} \right).$$

Implementing the transformation $T = T_1 \wedge T_2$, i.e.

$$T \left[\begin{array}{ccc} D_2, & \text{diameter}, & v_{21} \\ & \text{length}, & v_{22} \end{array} \right] = \left[\begin{array}{ccc} D, & \text{diameter}, & 10 \text{ cm} \\ & \text{length}, & <15, 20> \text{cm} \end{array} \right].$$

In this way, an effective connection of two pieces of plywood can be guaranteed.

Example 5.2.5

The basic function of desk lamp D is to illuminate the desk. To make the desk lamp D possess the voice-control function and the light-adjusting function, the voice-control device and the light-adjustment device must be added to the desk lamp at the same time. That is, we must add the transformations of the functions of the desk lamp at the same time.

$$T_1 \begin{bmatrix} \text{Illuminate, dominating object, desk} \\ \text{tools,} \qquad \text{desk lamp } D \end{bmatrix}$$

$$= \begin{bmatrix} \text{Illuminate, dominating object, desk} \\ \text{tools,} \qquad \text{desk lamp } D \end{bmatrix} \oplus \begin{bmatrix} \text{Control, dominating object, switch} \\ \text{tools,} \qquad \text{device } D_1 \\ \text{means,} \qquad \text{voice control} \end{bmatrix}$$

$$= \begin{bmatrix} \text{Illuminate} \oplus \text{Control, dominating object, desk} \oplus \text{switch} \\ \text{tools,} \qquad \text{desk lamp } D \oplus \text{device } D_1 \\ \text{means,} \qquad a \oplus \text{voice control} \end{bmatrix}$$

and

$$T_2 \begin{bmatrix} \text{Illuminate, dominating object, desk} \\ \text{tools,} \qquad \text{desk lamp } D \end{bmatrix}$$

$$= \begin{bmatrix} \text{Illuminate, dominating object, desk} \\ \text{tools,} \qquad \text{desk lamp } D \end{bmatrix} \oplus \begin{bmatrix} \text{Adjust, dominating object, light} \\ \text{tools,} \qquad \text{device } D_2 \\ \text{means,} \qquad \text{hand-rotary} \end{bmatrix}$$

$$= \begin{bmatrix} \text{Illuminate} \oplus \text{Control, dominating object, desk} \oplus \text{switch} \\ \text{tools,} \qquad \text{desk lamp } D \oplus \text{device } D_1 \\ \text{means,} \qquad a \oplus \text{hand-rotary} \end{bmatrix}.$$

So, $T = T_1 \wedge T_2$ is defined as

$$T \begin{bmatrix} \text{Illuminate, dominating object, desk} \\ \text{tools,} \qquad \text{desk lamp } D \end{bmatrix}$$

$$= \begin{bmatrix} \text{Illuminate} \oplus \text{Control} \oplus \text{Adjust, dominating object, desk} \oplus \text{switch} \oplus \text{light} \\ \text{tools,} \qquad \text{desk lamp } D \oplus \text{device } D_1 \oplus \text{device } D_2 \\ \text{means,} \qquad a \oplus \text{voice control} \oplus \text{hand-rotary} \end{bmatrix}$$

where the lighting mode of lamp D can be set randomly. Thus, it is denoted with a.

5.2.3 OR Transformation Method

If there exists either transformation T_1 or transformation T_2 such that $T_1 B_1 = B'_1$, $T_2 B_2 = B'_2$, and

$$T_1 B_1 \vee T_2 B_2 = B'_1 \vee B'_2 = B',$$

then the transformation $T = T_1 \vee T_2$ is known as the OR transformation of the transformations T_1 and T_2. The OR transformation involves the implementation of one transformation out of two or more available transformations.

For example, to contact someone you can choose to communicate face-to-face, by using a telephone, by correspondence, or by using a network technology. These four transformations are jointly an OR transformation.

As another example of an OR transformation, different processing methods can cause a product to consume energy differently. Thus, more than one transformation can be designed when processing the product so that the implementation of any one of the transformations can play a role in meeting the need of the product quality.

Case Analysis:

Example 5.2.6

In the processing of steel plate D, one of the following transformations can be taken: T_1 or T_2, denoted as $T = T_1 \vee T_2$, and different process engineers may choose different transformations:

$$T_1 = \begin{bmatrix} \text{Process,} & \text{dominating object,} & \text{steel plate } D \\ & \text{craft,} & \text{turning} \\ & \text{tools,} & \text{diamond} \end{bmatrix} = \begin{bmatrix} \text{Process,} & \text{dominating object,} & \text{steel plate } D \\ & \text{craft,} & \text{milling} \\ & \text{tools,} & \text{tungsten carbide} \end{bmatrix}$$

$$T_2 = \begin{bmatrix} \text{Process,} & \text{dominating object,} & \text{steel plate } D \\ & \text{craft,} & \text{turning} \\ & \text{lot sizing,} & \text{thousand} \end{bmatrix} = \begin{bmatrix} \text{Process,} & \text{dominating object,} & \text{steel plate } D \\ & \text{craft,} & \text{electrochemical corrosion} \\ & \text{lot sizing,} & \text{ten thousand} \end{bmatrix}.$$

Example 5.2.7

Innovating the desk lamp D by changing the value of the color of the desk lamp, by transforming the value of the shape of the desk lamp, or by transforming the value of the material of the desk lamp can all lead to new creative ideas for the product. These three transformations are jointly an example of the OR transformation, denoted as:

$$T_1 M_1 = T_1 \left(D, \quad \text{color}, \quad \text{white} \right) = \left(D, \quad \text{color}, \quad \text{colors} \right) = M_1'$$

$$T_2 M_2 = T_2 \left(D, \quad \text{shape}, \quad \text{cuboid} \right) = \left(D'', \quad \text{shape}, \quad \text{globe} \right) = M_2'$$

$$T_3 M_3 = T_3 \left(D'', \quad \text{material}, \quad \text{plastic} \right) = \left(D''', \quad \text{material}, \quad \text{glass} \right) = M_3'$$

where $T = T_1 \vee T_2 \vee T_3$ enables $T\left(M_1 \vee M_2 \vee M_3 \right) = M_1' \vee M_2' \vee M_3'$.

5.2.4 INVERSE Transformation Method

If there exists a transformation T such that $TB_0 = B_1$ and another transformation T', such that $T'B_1 = B_0$, that is, if $TT' = e$ satisfies $TB_0 = T\left(T'B_1\right) = eB_1 = B_1$, then T' is called the INVERSE transformation of T, denoted as $T^{-1} = T'$. The INVERSE transformation involves applying a transformation again for the transformed object so that this transformed object returns to its original form. It is a mode of reverse thinking.

Case Analysis:

Example 5.2.8

Child D played outside in the grass and brought home a small caterpillar in his hand. His mother, who was afraid of caterpillars, did not want her child to know her fear. So she asked her child to take the caterpillar back outside, otherwise its mother would worry about its whereabouts! After hearing this, the child took the caterpillar outside. After a while, the child came back in with a big caterpillar and a small caterpillar in his hands. The child said, "Its mother won't worry now, because I brought both of them inside."

Let us analyze the process of solving the contradictory problem facing the mother and the child in this case. Suppose D_1 = the small caterpillar, D_2 = the mother caterpillar that is supposed to be the mother. Let $M_1 = \left(D_1, \text{position}, \text{grass} \right)$ and $M_2 = \left(D_2, \text{position}, \text{grass} \right)$.

It can be found that the mother caterpillar thought the small caterpillar was transformed,

$$T_1 M_1 = \left(D_1, \quad \text{position}, \quad \text{home of } D \right) = M_1',$$

and that it would worry the mother caterpillar. To resolve the contradictory problem, the INVERSE transformation T_1^{-1} of T_1 must be applied so that

$$T_1^{-1} M_1' = T_1^{-1} \left(D_1, \quad \text{position}, \quad \text{home of } D \right) = \left(D_1, \quad \text{position}, \quad \text{grass} \right) = M_1.$$

The child believed that the transformation T_2 must be implemented on M_2 to solve the contradiction between M_1' and M_2 so that

$$T_2 M_2 = T_2 \left(D_2, \quad \text{position}, \quad \text{grass} \right) = \left(D_2, \quad \text{position}, \quad \text{home of } D \right) = M_2'.$$

In this way, the goal of not worrying the caterpillar's mother can be achieved. The child used the PRODUCT transformation $T = T_2\,T_1$, which solved the contradictory problem.

Example 5.2.9

A young engineer is building a rocket. He is seeking a solution for a balance problem she is observing during the rocket's launch. He believes the issue is related to the weight of the rocket as it rises during launch. He had been looking for a "heat-resistant" material to attach to the bottom of the rocket, but he has found no acceptable material that allowed for successful launch. He decided to try the innovative solution of using a wood material that is not heat-resistant to attach to the bottom of the rocket. The burning of the wood material helps reduce the weight of the rocket, and the rocket achieves to the engineer's desired height and speed. That is, when t stands for an arbitrary time moment, we have the following matter-element:

$$M(t) = \begin{bmatrix} \text{Material } D(t), & \text{effect}, & \text{balance rocket}(t) \\ & \text{characteristic}, & \text{not heat-resistant}(t) \\ & \text{weight}, & v(t) \end{bmatrix} = \begin{bmatrix} M_1(t) \\ M_2(t) \\ M_3(t) \end{bmatrix}.$$

If he want to implement the following transformation

$$T_1 M_2(t) = T_1 \left(\text{Materail } D(t), \quad \text{characteristic}, \quad \text{not heat-resistant}(t) \right)$$

$$= \left(\text{Materail } D'(t), \quad \text{characteristic}, \quad \text{heat-resistant}(t) \right) = M_2'(t)$$

so that

$$T_1 M(t) = \begin{bmatrix} M_1(t) \\ M_2'(t) \\ M_3(t) \end{bmatrix} = M'(t),$$

but it is impossible to find out the material $D'(t)$.

When $t = t_1 = $ "an early stage of launching the rocket", the following transformation can be implemented:

$$T_1^{-1} M_2'(t_1) = T_1^{-1} \left(\text{Material } D'(t_1), \quad \text{characteristic}, \quad \text{heat-resistant}(t_1) \right)$$

$$= \left(\text{Material } D(t_1), \quad \text{characteristic}, \quad \text{not heat-resistant}(t_1) \right) = M_2(t_1).$$

In other words:

$$T_1^{-1}M'(t_1) = T_1^{-1} \begin{bmatrix} \text{Material } D'(t_1), & \text{effect,} & \text{balance rocket}(t_1) \\ & \text{characteristic,} & \text{heat-resistant}(t_1) \\ & \text{weight,} & v(t_1) \end{bmatrix}$$

$$= \begin{bmatrix} \text{Material } D(t_1), & \text{effect,} & \text{balance rocket}(t_1) \\ & \text{characteristic,} & \text{not heat-resistant}(t_1) \\ & \text{weight,} & v(t_1) \end{bmatrix} = \begin{bmatrix} M_1(t_1) \\ M_2(t_1) \\ M_3(t_1) \end{bmatrix} = M(t_1).$$

When $t = t_2 =$ "the rocket launching to a certain height", the following transformation is implemented:

$$T_1^{-1}M_2'(t_2) \wedge T_2 M_1(t_2) \wedge T_3 M_3(t_2) = \begin{bmatrix} \text{Material } D(t_2), & \text{effect,} & 0 \\ & \text{characteristic,} & \text{not hot-resistant}(t_2) \\ & \text{weight,} & 0 \end{bmatrix}$$

$$= \begin{bmatrix} M_1'(t_2) \\ M_2(t_2) \\ M_3'(t_2) \end{bmatrix} = M''(t_2).$$

Thus we only need to select a wood material as the material $D(t)$ instead of using a heat-resistant material.

5.3 Conductive Transformation Methods

QUESTIONS AND THINKING:

■ What is the butterfly effect?
■ Why does the statement that "a change of a tiny part may affect the whole" mean?
■ What is a "virtuous circle"? What is a "vicious circle"?
■ Can these questions be described by using a formalized and quantitative method?
■ How can we use these descritpions to innovate or solve contradictory problems?

Due to the universality of interrelations and implications among things, one transformation of an object will lead to transformations of other relevant objects—this is called a conductive transformation. Each conductive transformation is based on the correlations that exist among objects. Therefore, relevant analysis of a certain object should be conducted with the use of a correlation network before implementing an extension transformation on the object.

The so-called conductive transformation method refers to the way that people consciously use a conductive transformation to innovate or to solve contradictory problems. Before introducing the method, the formal definition, types of conductive transformations, and conductive effects are introduced.

5.3.1 Definition of Conductive Transformations

For basic-elements B_1 and B_2, if $B_1 \sim B_2$, then corresponding to the active transformation $\varphi B_1 = B_1'$ implemented on B_1, there must be a passive transformation $T_\varphi B_2 = B_2'$ for B_2. With $\varphi \Rightarrow T_\varphi$, T_φ is called a first-order conductive transformation caused by φ, or a conductive transformation for short.

There must be an implication relationship between the active transformation and its conductive transformation. If there are multiple conductive transformations occurring, a set of conductive transformations will be formed.

For example, because the hardness of a gear is correlated to its life, that is

$$\Big(\text{Gear } D, \quad \text{hardness}, \quad v_1\Big) \sim \Big(\text{Gear } D, \quad \text{life}, \quad v_2\Big).$$

If an active transformation is implemented on the hardness's value of the gear, then it will cause a conductive transformation to be conducted on the service life's value of the gear. That is,

$$\varphi\Big(\text{Gear } D, \quad \text{hardness}, \quad v_1\Big) = \Big(\text{Gear } D' \quad \text{hardness}, \quad v_1'\Big),$$

$$T_\varphi\Big(\text{Gear } D, \quad \text{life}, \quad v_2\Big) = \Big(\text{Gear } D' \quad \text{life}, \quad v_2'\Big),$$

and $\varphi \Rightarrow T_\varphi$.

5.3.2 Types of Conductive Transformations

Because there is a close relationship among the three elements of a basic-element $(O, \quad c, \quad v)$, especially among the three elements of a matter-element, a transformation of one element may lead to changes of the other elements, which is called the conductive transformation among elements of the basic-element.

According to the correlative network method introduced in Chapter 3, correlations among basic-elements can be divided into correlations of different characteristics with the same object, correlations of the same characteristics with different objects, and correlations of different characteristics with different objects. Therefore, conductive transformations among basic-elements also include basic-element conductive transformations with the same object, basic-element conductive transformations with different objects, and other complex conductive transformations. A variety of basic conductive transformations are introduced below.

5.3.2.1 Conductive Transformations Among Elements of a Basic-element

Suppose that basic-element $B(t) = (O(t), c, v(t))$ is given, satisfying $v(t) = c(O(t))$, and the object $O(t)$ is implemented with an active transformation. The conductive transformation of its value about the characteristic c may also occur. Similarly, because the active transformation is implemented on the value $v(t)$ of the basic element, its object $O(t)$ may also experience a conductive transformation. A special case occurs no conductive transformation appears.

There are various forms of conductive transformations, mainly including:

1. If $\varphi O(t) = O'(t)$, it must follow that $\varphi \Rightarrow T_\varphi$, satisfying $T_\varphi v(t) = v'(t)$. When the active and conductive transformations are unspecified implicitly, the active transformation and conductive transformation inside the basic-element are represented with the following transformation:

$$TB(t) = T\big(O(t), \quad c, \quad v(t)\big) = \big(O'(t), \quad c, \quad v'(t)\big).$$

When the consideration of parameters is unnecessary, this can be simplified as:

$$TB = T\big(O, \quad c, \quad v\big) = \big(O', \quad c, \quad v'\big).$$

For example, if a table is replaced by a chair, the value of "length" must be changed. The occurrence of the same value only represents an exceptional case. That is

$$T\big(\text{Table } O, \quad \text{length}, \quad 1 \text{ m}\big) = \big(\text{Chair } O', \quad \text{length}, \quad 0.5 \text{ m}\big).$$

2. If $\varphi v(t) = v'(t)$, it must follow that $\varphi \Rightarrow T_\varphi$, satisfying $T_\varphi O(t) = O'(t)$. When the active and conductive transformations are unspecified implicitly, the active transformation and conductive transformation inside the basic-element are represented with the following transformation:

$$TB(t) = T\big(O(t), \quad c, \quad v(t)\big) = \big(O'(t), \quad c, \quad v'(t)\big).$$

When the consideration of parameters is unnecessaryand will not leading to confusion, it can be simplified as:

$$TB = T(O, \quad c, \quad v) = (O', \quad c, \quad v').$$

For example, if the color of a table lamp is white, and if white color is changed to a red one, it must be another table lamp. That is:

$$T(\text{Table lamp } O, \quad \text{color}, \quad \text{white}) = (\text{Table lamp } O', \quad \text{color}, \quad \text{red}).$$

5.3.2.2 Conductive Transformation on Basic-elements with the Same Object

Suppose that $B_1 = (O, c_1, v_1), B_2 = (O, c_2, v_1)$, and $B_1 \sim B_2$. If the active transformation $\varphi B_1 = B_1'$ is implemented, then there must exist a conductive transformation $T_\varphi B_2 = B_2'$ of the basic-element with the same object. If various kinds of correlative networks are formed by multiple basic-elements, then the conductive transformation will have many different forms.

Case Analysis:

Example 5.3.1

Changing the material of a light fixture D may result in changes in the values of characteristics such as weight, craft, cost, price, sales volume, profit, etc. Changing the values of length, width, and height of the light fixture may similarly result inchanges in the values of volume, weight, capacity, etc.

These are the conductive transformations among basic-elements of different characteristics with the same object based on the following correlative net (tree) of basic-elements:

$$M_1 = (D, \quad \text{material}, \quad v_1) \sim \begin{cases} M_2 = (D, \quad \text{weight}, \quad v_2) \\ M_3 = (D, \quad \text{craft}, \quad v_3) \\ M_4 = (D, \quad \text{cost}, \quad v_4) \\ M_5 = (D, \quad \text{price}, \quad v_5) \\ M_6 = (D, \quad \text{sales volume}, \quad v_6) \\ M_7 = (D, \quad \text{profit}, \quad v_7) \end{cases}.$$

This correlative network tells us that if an active transformation φ is implemented on M_1, it will lead $M_i (i = 2, \cdots, 7)$ to experience conductive transformation $T_{\varphi i}$, that is, $\varphi \Rightarrow \bigwedge\limits_{i=2}^{7} T_{\varphi i}$.

Similarly, from

$$M_8 \vee M_9 \vee M_{10} = \left(D, \quad \text{length}, \quad v_8\right) \vee \left(D, \quad \text{width}, \quad v_9\right) \vee \left(D, \quad \text{height}, \quad v_{10}\right)$$

$$\hat{\sim} \begin{cases} \left(D, \quad \text{weight}, \quad v_2\right) = M_2 \\ \left(D, \quad \text{volume}, \quad v_{11}\right) = M_{11} \\ \left(D, \quad \text{capacity}, \quad v_{12}\right) = M_{12} \end{cases}$$

we have $M_8 \vee M_9 \vee M_{10} \sim M_2 \wedge M_{11} \wedge M_{12}$. If the active transformation φ_i is conducted on any of $M_i (i = 8, 9, 10)$, it will lead $M_i (i = 2, 11, 12)$ to experience conductive transformation $T_{\varphi i}$ at the same time: $\bigvee\limits_{i=8}^{10} \varphi_i \Rightarrow T'_{\varphi 2} \wedge T_{\varphi 11} \wedge T_{\varphi 12}$.

5.3.2.3 Conductive Transformations on Basic-elements with Different Objects

Suppose that $B_1 = (O, \quad c_1, \quad v_1), B_2 = (O, \quad c_2, \quad v_1)$, and $B_1 \sim B_2$ are given. If the active transformation $\varphi B_1 = B'_1$ is implemented, there must exist conductive transformation $T_\varphi B_2 = B'_2$ of the basic-element with different objects. If various kinds of correlative networks are formed by multiple basic-elements, then the conductive transformation will have many different forms.

Case Analysis:

Example 5.3.2

A change of the value of the lighting color will lead to a change of the value of users' moods. A change of the value of the lighting structure will lead to a change of the value of users' usage modes. These are conductive transformations among basic-elements of different characteristics with different objects, which are based on the following correlative network (tree) of basic-elements:

$$M_1 = \left(\text{Lighting-fixture } D_1, \quad \text{color}, \quad v_1\right) \sim M_2 = \left(\text{User } D_2, \quad \text{mood}, \quad v_2\right),$$

$$M_3 = \left(\text{Lighting-fixture } D_1, \quad \text{structure}, \quad v_3\right) \sim M_4 = \left(\text{User } D_2, \quad \text{using-way}, \quad v_4\right).$$

If an active transformation φ_1 is implemented on M_1, it will cause the conductive transformation $T_{\varphi 1} M_2 = M'_2$ to be conducted on M_2. If an active transformation φ_3 is implemented on M_3, it will cause the conductive transformation $T_{\varphi 3} M_4 = M'_4$ to be conducted on M_4.

Example 5.3.3

The heights of a desk and a chair are correlative to the user's height. Desks of different heights should be equipped with chairs of different heights, and users of different heights should use tables and chairs of different heights. These are conductive transformations of a same characteristic with different objects, which are based on the following correlative network (tree) of basic-elements:

$$M_1 = \left(\text{Desk } D_1, \quad \text{height}, \quad v_1\right) \sim M_2 = \left(\text{Chair } D_2, \quad \text{height}, \quad v_2\right),$$

$$M_3 = \left(\text{User } D_3, \quad \text{height}, \quad v_3\right) \sim M_1 = \left(\text{Desk } D_1, \quad \text{height}, \quad v_1\right).$$

If $\varphi_1 M_1 = (\text{Desk } D_1', \quad \text{height}, \quad v_1') = M_1'$ is implemented, there must be a conductive transformation $T_{\varphi 1}$ such that

$$T_{\varphi 1} M_2 = \left(\text{Chair } D_2', \quad \text{height}, \quad v_2'\right) = M_2'.$$

If $\varphi_3 M_3 = (\text{User } D_3', \quad \text{height}, \quad v_3') = M_3'$ is implemented, there must be a conductive transformation $T_{\varphi 3}$ such that

$$T_{\varphi 3} M_1 = \left(\text{Desk } D_1'', \quad \text{height}, \quad v_1''\right) = M_1''.$$

Also, there will be a conductive transformation working on M_2. All details are omitted here.

5.3.2.4 Complex Conductive Transformations

In many practical cases, there is both a correlation of basic elements with the same object and a correlation of basic elements with different objects in a correlative network—an AND correlation and an OR correlation. Therefore, the implementation of an active transformation may lead to a complex conductive transformation.

For example, a certain class of a college is preparing for its thirtieth graduation anniversary party. The organizer established an online chat group to invite all classmates to join the celebration. However, many classmates didn't know how to use this chat application. Due to the attraction of the planned classmate reunion, they joined the chat group with the assistance of their children or friends. Those who joined the group increased their daily use time. Accordingly, the use of mobile phone traffic also increased. Furthermore, the intimating degrees among classmates have been increased. This is a complex conductive transformation.

Complex conductive transformations exist in all areas of life. If we can carefully analyze and make use of conductive transformations, it will greatly benefit innovative solutions to contradictory problems. Detailed discussion will not be introduced here.

5.3.2.5 n Times of Conductive Transformations

If an active transformation is implemented, a series of chain reactions will occur, and such transformations are called *n* times of conductive transformations. If the active transformation is represented by φ, and the *n*th conductive transformation is represented by $_{n-1}T_n$, then there is an implication relation of the following transformations:

$$\varphi \Rightarrow {}_\varphi T_1 \Rightarrow {}_1 T_2 \Rightarrow \cdots \Rightarrow {}_{n-2}T_{n-1} \Rightarrow {}_{n-1}T_n.$$

Thus, a conductive transformation chain is formed by these conductive transformations.

Case Analysis:

Example 5.3.4

A change in the value of the transmission ratio of transmission *F* for a certain car *D* will result in a change in the value of the fuel consumption of engine *E*, and the change of the value in the fuel consumption of engine *E* will lead to a change in the value of use time of the car *D*. In other words:

$$\left(F, \quad \text{transmission ratio,} \quad v_1 \right) \sim \left(E, \quad \text{fuel consumption,} \quad v_2 \right) \sim \left(D, \quad \text{using time,} \quad v_3 \right).$$

Thus, we have

$$\varphi\left(F, \quad \text{transmission ratio,} \quad v_1 \right) = \left(F, \quad \text{transmission ratio,} \quad v_1' \right)$$
$$_\varphi T_1\left(E, \quad \text{fuel consumption,} \quad v_2 \right) = \left(E, \quad \text{fuel consumption,} \quad v_2' \right),$$
$$_1 T_2\left(D, \quad \text{using time,} \quad v_3 \right) = \left(D, \quad \text{using time,} \quad v_3' \right)$$

which can be shown as a secondary conductive transformation: $\varphi \Rightarrow {}_\varphi T_1 \Rightarrow {}_1 T_2$.

In particular, the *n*th conductive transformation may also cause a conductive transformation in the basic-element that was implemented with the initial active transformation. That is, a ring of conductive transformations may be formed. Detailed discussion of this is not given here. Readers who are interested in it can refer to the monograph ***Extenics*** for more details.

The concept of *n* times of conductive transformations is a formalized expression of what we usually refer to as "virtuous circles" or "vicious circles." Readers who are interested in it can refer to the monograph ***Extenics*** for more details.

5.3.3 Conductive Effect

The so-called conductive effect is an important index for the quantitative study of conductive transformations. It can be studied by selecting an appropriate perspective. In the following, a computational method of conductive effects is given with a matter-element as an example.

Matter-elements B_0 and B are given. If $B_0 = (O_0,\ c_0,\ v_0)$ and $B = (O,\ c,\ v)$ with $B_0 \sim B$, and if an active transformation φ exists such that $\varphi B_0 = (O_0',\ c_0,\ v_0') = B_0'$, and if there is a conductive transformation $T_\varphi B = (O',\ c,\ v') = B'$, the first-order conductive effect of the basic-element B for φ on characteristic c is $c(T_\varphi) = v' - v$. The active variable of φ on c_0 is $c_0(\varphi) = v_0' - v_0$.

If $c(T_\varphi) > 0$, the effect is called a positive conductive effect on characteristic c; if $c(T_\varphi) < 0$, the effect is called a negative conductive effect on characteristic c. If $c(T_\varphi) = 0$, it is said that the conductive transformation is a "none conductive effect" on characteristic c. The ratio

$$\gamma = \frac{c(T_\varphi)}{|c_0(\varphi)|}$$

is called the conductive degree of the conductive transformation T_φ on the active transformation φ.

In addition, the first-order nth conductive effect and m-order conductive effect are more complicated. This is not discussed in detail here; interested readers may refer to the monograph **Extenics** for more in-depth study.

In general, if $B_0 \sim \bigwedge\limits_{i=1}^{n} B_i, c(_\varphi T_i) = c(B_i') - c(B_i)$,

$$c(T_\varphi^{(1)}) = \sum_{i=1}^{n} c(_\varphi T_i) = \sum_{i=1}^{n} \left[c(B_i') - c(B_i) \right]$$

is called comprehensive first-order first-time conductive effect. The first-order nth conductive effect is denoted as $c(T_\varphi^{(n)})$.

Case Analysis:

Example 5.3.5

A change in the output of a car company D_1 will result in a change in the price of car D_2; in turn, it will lead to a change in the sales volume of the car company. Furthermore, it will result in a change in the volume of deposits of users D_3 who buy cars:

$$\varphi\Big(D_1,\quad \text{output},\quad 10{,}000\ \text{cars/year}\Big) = \Big(D_1',\quad \text{output},\quad 15{,}000\ \text{cars/year}\Big)$$

$$_\varphi T_1\Big(D_2,\quad \text{price},\quad 200\ \text{thousand yuan}\Big) = \Big(D_2',\quad \text{price},\quad 190\ \text{thousand yuan}\Big)$$

$$_1 T_2\Big(D_1,\quad \text{sales volume},\quad 9{,}000\ \text{cars/year}\Big) = \Big(D_1',\quad \text{sales volume},\quad 13{,}000\ \text{cars/year}\Big)$$

$$_2 T_3\Big(D_3,\quad \text{volume of deposit},\quad 300\ \text{thousand yuan}\Big) = \Big(D_3,\quad \text{volume of deposit},\quad 150\ \text{thousand yuan}\Big)$$

So, the first-order third conductive effect on the characteristic "volume of deposits" is $c(T_\varphi) = 15 - 30 = -15$.

5.3.4 General Steps for the Conductive Transformation Method

Each conductive transformation is a passive transformation that is the result of the implementation of one or more active transformations based on the correlation rule described in Chapter 3.

The conductive transformation method is to innovate or solve contradictory problems by actively making use of conductive transformations. The general steps of this method are shown here:

1. Determine whether the active transformation can solve the problem. If not, the next step can proceed directly; if it does, judge whether the transformation time or effect is suitable for the transformation, or estimate the cost of the transformation. If the transformation time and/or effect is appropriate, then end; if not, then go to next step.
2. Analyze the basic-element B_1 on which the active transformation was implemented. Form a correlative network to judge whether the active transformation was implemented with its relevant basic-element. If yes, judge whether the required conductive transformation can be implemented on the basic-element B_1; if yes, the problem will be solved.

There are numerous methods for implementing conductive transformations. These methods can be divided into a first-order conductive transformation method and a multi-order conductive transformation method according to the division of conduction order. Conductive transformation methods can be divided into that of conductive transformations caused by transformations of basic-elements, that of conductive transformations caused by rule transformations, that of conductive transformations caused by domain transformations, that of conjugate conductive transformations, etc., according to different objects of the active transformations.

Note: The conductive transformation method may also transform an original contradictory problem into a non-contradictory problem while generating a new conductive transformation for other basic-elements, resulting in the generation of a new contradictory problem. At this point, a new transformation must also be undertaken to solve the new contradictory problem.

Case Analysis:

Example 5.3.6

If the vehicle mass is increased by 20 kg, the fuel consumption will be increased by 1%. This is a hard problem to break through in a short time for engine designs. Suppose

$$M_1 = \left(\text{Engine } D_1, \quad \text{fuel consumption}, \quad 1.21 \text{ L/km} \right)$$
$$M_2 = \left(\text{Car } D_2, \quad \text{quality}, \quad a \text{ kg} \right).$$

If the following active transformation is made directly,

$$T_1 M_1 = T_1 \left(\text{Engine } D_1, \quad \text{fuel consumption}, \quad 1.21 \text{ L/km} \right)$$

$$= \left(\text{Engine } D_1', \quad \text{fuel consumption}, \quad 1.20 \text{ L/km} \right),$$

it can been seen that it is hard to realize the desired decrease of 1% of fuel consumption of the engine. However, because $M_2 \sim M_1$, through the implementation of the transformation

$$T_2 M_2 = T_2 \left(\text{Car } D_2, \quad \text{quality}, \quad a \text{ kg} \right) = \left(\text{Car } D_2', \quad \text{quality}, \quad (a - 20) \text{ kg} \right),$$

transformation T_1 can be realized, i.e. $T_2 \Rightarrow T_1$, which can also realize the target of decreasing the fuel consumption by 1%.

This is the method adopted by a Japanese automobile designer to deal with the issue of energy-saving vehicles. The company is using the principle of cutting the weight of car body for achieving the reduction in energy consumption. The example also tells us that when a target cannot be achieved with the active transformation, you can consider using the conductive transformation method.

5.4 Conjugate Transformation Methods

QUESTIONS AND THINKING:

■ Why do companies spend money on advertising?
■ Why do some companies turn losses into profits by changing the leadership team?
■ Why should defective products be recalled?
■ Why do companies build canteens, kindergartens, activity rooms, etc. for the staff?

A conjugate transformation stands for a special type of conductive transformation based on the conjugate rule and the conjugate analysis method of matters. The active transformation of the conjugate part of the matter is introduced before introducing the conjugate transformation.

5.4.1 Transformations of Conjugate Parts

The so-called conjugate part transformation refers to the active transformation of any part in the four conjugate pairs of matters. For example, in terms of products, transformations of the values of shape, size, and raw materials, etc., are transformations of the material parts of a product, while transformations of the values of a brand name and popularity are transformations of the nonmaterial parts of a product. Transformations of the composition of a product are transformations of the hard parts of a product, while transformations of the structure of a product and the connecting mode of each component are transformations of the soft parts of a product. The development of potential functions of the apparent part of a product, such as designing the product packaging for reuse is a transformation of the apparent part of a product, while modifications of the parts that are disadvantageous to consumers, such as minimizing the negative effect of drugs on humans and reducing the radiation of mobile phones are transformations of the negative part of a product.

According to the conjugation rules in Chapter 4, generally and without considering the mid-part, the matter O_m can be divided by materiality, systematicness, dynamism, and antagonism into four conjugate parts, i.e. the material part $re(O_m)$ and nonmaterial part $im(O_m)$, the soft part $sf(O_m)$ and the hard part $hr(O_m)$, the latent part $lt(O_m)$ and the apparent part $ap(O_m)$, and the negative part $ng_c(O_m)$ and the positive part $ps_c(O_m)$. Any transformation of a certain part of these eight parts is called a conjugate part transformation.

Transformations of basic-elements formed by each conjugate part are correspondingly denoted as:

$$T_{im}M_{im} = M'_{im}, \quad T_{re}M_{re} = M'_{re},$$

$$T_{sf}M_{sf} = M'_{sf}, \quad T_{hr}M_{hr} = M'_{hr},$$

$$T_{lt}M_{lt} = M'_{lt}, \quad T_{ap}M_{ap} = M'_{ap},$$

$$T_{ng_c}M_{ng_c} = M'_{ng_c}, \quad T_{ps_c}M_{ps_c} = M'_{ps_c}.$$

Research on transformations of the conjugate part of a matter is the basis for studying conjugate transformations. The transformation mode of the conjugate part is consistent with that of the matter, including basic transformations and operations of transformations.

5.4.2 Conjugate Transformation Rules

Because various correlations exist among conjugate parts, an active transformation of the material part may cause a conductive transformation to occur in the values of its nonmaterial part about multiple characteristics. This applies in the other conjugate parts as well. If the property can be fully utilized, the effect of "serving multiple purposes" can be achieved. In fact, there are other conductive transformations inside the matter. For example, one transformation of a part in the material part may lead to a transformation of other relevant parts in the material part. Likewise, one transformation of a part in the nonmaterial part may lead to a transformation of the other relevant parts in the nonmaterial part.

According to Conjugate Rule 4.1.2, one transformation of a certain conjugate part will lead to one transformation of another conjugate part in the same conjugate pair, which is called aconjugate transformation. Conjugate transformations can be divided into nonmaterial and material conjugate transformations, soft and hard conjugate transformations, latent and apparent conjugate transformations, and negative and positive conjugate transformations. Conjugate transformations are governed by conjugate transformation rules.

Conjugate transformation rule 5.4.1 One transformation of the material part of the matter will lead to one conductive transformation to occur in its relevant non-material part. Conversely, one transformation of the nonmaterial part of the matter will result in one conductive transformation to occur in its relevant material part.

Let M_{re} be the material part of a certain matter O_m, M_{im} the nonmaterial part of the matter. If $T_{im}M_{im} = M'_{im}$, it must exist that $_{im}T_{re}$ and $T_{im} \Rightarrow {_{im}T_{re}}$. With $_{im}T_{re}M_{re} = M'_{re}$, $_{im}T_{re}$ is called a material and nonmaterial conjugate transformation of T_{im}. Similarly, $_{re}T_{im}$ is called a material and nonmaterial conjugate transformation of T_{re}, where T_{re} and T_{im} represent active transformations of the material part M_{re} and the nonmaterial part M_{im} of the matter O_m, respectively, and $_{re}T_{im}$ and $_{im}T_{re}$ represent conductive transformations of the material part M_{im} and the nonmaterial part M_{re} of the matter O_m, respectively.

Case Analysis:

Example 5.4.1

For a personal computer, all the accessories, display, audio, and connecting lines within computer are the material parts. Not only all do accessories have to be connected with plugs and slots on the basis of technical requirements while assembling the computer, but an operating system and various kinds of applications must also be installed for the computer to operate normally. All physical parts of the computer belong to the material part. Its nonmaterial parts include the computer's brand value, its appearance, its popularity, its reputation, and how user-friendly the operating system and various applications are. To change the nonmaterial parts, such as improving the brand value, the company must invest enough research and material resources to change the function and quality of the hardware, or they may carry out a lot of advertisingto achieve these changes. These transformations belong to the category of material and nonmaterial conjugate transformations.

Material and nonmaterial conjugate transformations are often adopted in the process of solving contradictory problems. However, the realization of material and nonmaterial conjugate transformations may not always be conductive to the development of enterprises. When a negative effect is generated from a conductive transformation among the conjugate parts, it will hinder the development of the enterprise.

For example, suppose an enterprise spends huge sums of money on advertisement to improve its reputation. Although its popularity has increased greatly in a short period of time, the company has invested so much money on advertising that it fails to invest enough money in production and product development. That is, the transformation in the nonmaterial part leads to a decline in the actual strength of the enterprise, eventually causing the enterprise's reputation to deteriorate. Therefore, we must pay great attention to the conductive effects and take effective measures to ensure that the employed transformation to become a development-driving force. That should also be noted in solving contradictory problems.

Example 5.4.2

The batteries of laser pen, the laser diode, laser components, housings, indicator lights, etc., are material parts of a laser pointer, while the indicator light emitted by the laser pointer, the figures that the laser pointer illuminates, and the brand are nonmaterial parts.

To change the shape and the size of the figures illuminated by the laser pointer, the material and nonmaterial conjugate pair method must first be utilized to identify its material parts, such as the battery, the laser diode, and the laser components, etc. Then, according to the conjugate transformation rule, one transformation of a material part will lead to one conductive transformation of its relevant nonmaterial part(s). That is, the material and nonmaterial conjugate transformations thus change the shape and the size of the figure illuminated by the laser pointer.

Conjugate transformation rule 5.4.2 One transformation of the hard part of the matter will lead to one conductive transformation in its relevant soft part. Conversely, one transformation of the soft part of the matter will result in one conductive transformation in its relevant hard part. The conductive transformations are respectively called soft and hard conjugate transformations that can be symbolized as $T_{hr} \Rightarrow {}_{hr}T_{sf}$, $T_{sf} \Rightarrow {}_{sf}T_{hr}$ where T_{hr} and T_{sf} represent the active transformations of the hard part M_{re} and the soft part M_{im} of the matter O_m, respectively, while ${}_{hr}T_{sf}$ and ${}_{sf}T_{hr}$ represent the conductive transformations of the soft part M_{sf} and the hard part M_{hr} of the matter O_m, respectively.

Case Analysis:

Example 5.4.3

Any organization that considers combining department staffs (hard part) must consider the coordination among the personnel (soft part) as well as the requirements of each position in the department. If the staff in the department work against each other and undermine each other's efforts, it will be difficult for each person to fully function at his or her potential. If the staff in the department work harmoniously together and coordinate with each other, strong cohesion and creative ideas will be generated. Therefore, changes in the internal relations of the department (soft part) may lead to changes in the functionality (hard part) of each person in the department. In addition, changes of external relations will also exert an effect on the hard part.

When recruiting staff, the employer does not look only at an applicant's situation, but also pays attention to his or her previous work. If the applicant previously engaged in relevant work, he or she will be prioritized during the recruitment. Because the applicant has relevant work experience, there will be a relevant well-connected network. Once this person is hired, his or her external relationships (soft part) will come to the new position with the applicant. Thus it can be seen that the changes in the hard part inevitably lead to changes in the soft part.

Example 5.4.4

Historically, many weak countries had to sign "unequal treaties" with great powers for peace with "ceding lands" for compromises. The practice of using the transformation of "land" (i.e. the hard part) to obtain an improvement of "the relation" (i.e. the soft part) and to realize peace is a method to obtain a soft transformation with the hard transformation.

Conjugate transformation rule 5.4.3 One transformation of the negative part of the matter will lead to one conductive transformation in its relevant positive part. Conversely, one transformation of the positive part of the matter will result in one conductive transformation in its relevant negative part. The conductive transformations are respectively called the negative and positive conjugate transformations, symbolized as $T_{\mathrm{ng}_c} \Rightarrow {}_{\mathrm{ng}_c}T_{\mathrm{ps}_c}$ and $T_{\mathrm{ps}_c} \Rightarrow {}_{\mathrm{ps}_c}T_{\mathrm{ng}_c}$, where T_{ng_c} and T_{ps_c} represent the active transformations of the negative part M_{ng_c} and the positive part M_{ps_c} of the matter O_m with respect to the characteristic c, respectively, and ${}_{\mathrm{ng}_c}T_{\mathrm{ps}_c}$ and ${}_{\mathrm{ps}_c}T_{\mathrm{ng}_c}$ respresent the conductive transformations of the negative part M_{ps_c} and the positive part M_{ng_c} of the matter O_m with respect to the characteristic c, respectively.

Case Analysis:

Example 5.4.5

Suppose an enterprise has unused parts of their plant and equipment after restructuring and adjusting their production structures. As for profits, these surplus "plant and equipment" have become a negative part of the enterprise because they cannot be used to generate benefits due to deteriorating values or the cost of maintenance and repair. To change this situation, the company must carefully plan on how to make these negative parts serve the enterprises' goals. According to the principle of negative and positive conjugate transformations, the negative part can be converted to the positive part through some transformations, such as renting out the surplus plants and equipment.

Example 5.4.6

The portion of vehicle exhaust that is not fully burned is not in compliance with the national emission standard. The exhaust return pipe in the exhaust processing system is used to re-combust the exhaust, thus reducing the content of toxic gas and creating heat energy. In this case, the incompletely burned exhaust is a negative aspect of the vehicle, but it is converted to a positive part—heat—through a transformation of "returning for re-combustion" on the negative part.

Conjugate transformation rule 5.4.4 One transformation of the latent part of the matter will lead to one conductive transformation in its relevant apparent part. Conversely, one transformation of the apparent part of the matter will result in one conductive transformation in its relevant latent part. The conductive transformations are respectively called a latent and apparent conjugate transformation, symbolized as $T_{\mathrm{lt}} \Rightarrow {}_{\mathrm{lt}}T_{\mathrm{ap}}$, $T_{\mathrm{ap}} \Rightarrow {}_{\mathrm{ap}}T_{\mathrm{lt}}$, where T_{lt} and T_{ap} represent the active transformations of the latent part M_{lt} and the apparent part M_{ap} of the matter O_m, respectively, and ${}_{\mathrm{lt}}T_{\mathrm{ap}}$ and ${}_{\mathrm{ap}}T_{\mathrm{lt}}$ represent T_{lt} and T_{ap}'s conductive transformations of the latent part M_{lt} and the apparent part M_{ap} of the matter O_m, respectively.

The latent part of a matter has both positive and negative latent parts. For example, "hidden dangers" in enterprises and "crises" implied in a development process are the negative latent parts of enterprises, while the "potential markets" of enterprises, the "potential productivity" of employees, and the "development potential" of enterprises are the positive latent parts of enterprise. Therefore, an important task for enterprises is to determine how to implement an effective transformation to make the positive latent part of the enterprise apparent as soon as possible, and to prevent the negative latent part from becoming apparent or instead to make it turn into a positive latent part.

Enterprises are often defeated by their potential competitors. Therefore, accurately identifying potential competitors is a key for enterprises to stay successful. For Eastman Kodak, most people think that its rival is Fujifilm. However, the biggest threat encountered by Kodak actually came from the rapid development of home video camera technology, i.e. digital cameras developed by Canon and Sony.

As another example, NBA star LeBron James announced that he would become a free agent by the end of 2013–2014 NBA season. At that time, no one was sure which team he would join, with many people considering that he was likely to return to his hometown team, the Cleveland Cavaliers. In other words, LeBron James was a latent part of the Cleveland Cavaliers. Some ticket vendors immediately purchased a large number of Cavaliers tickets for the next season. When James announced he would join Cavaliers, he became an apparent part of Cavaliers. As a result, the ticket prices increased several-fold, and the ticket sellers made a lot of money.

5.4.3 General Steps of Conjugate Transformation Method

The method of utilizing transformations of conjugate parts and conjugate transformations to innovate or solve contradictory problems is called the conjugate transformation method. Its general steps are similar to those of the conductive transformation method, which will be illustrated by using examples as follows.

Example 5.4.7

Mobile phone D of a certain brand with a medium level of system performance and a low degree of software innovation has a medium cost-effective public praise. It is sold at the price of 1,999 yuan. In such conditions, what could one do to increase the sales volume of the enterprise?

Because the situation involves material and nonmaterial conjugate parts, the material and nonmaterial conjugate analysis and transformation methods can be used.

First, let us conduct a material and nonmaterial conjugate analysis on the mobile phone D based on the given conditions:

$$
M_{im} = \begin{bmatrix} D, & \text{cost-effective,} & \text{medium} \\ & \text{public praise,} & \text{medium} \\ & \text{price,} & \text{1,999 yuan} \\ & \text{system performance,} & \text{medium} \\ & \text{degree of software innovation,} & \text{low} \end{bmatrix} = \begin{bmatrix} D, & c_1, & \text{medium} \\ & c_2, & \text{public} \\ & c_3, & \text{1,999 yuan} \\ & c_4, & \text{medium} \\ & c_5, & \text{low} \end{bmatrix}
$$

$$
M_{re} = \begin{bmatrix} D, & \text{level of hardware configuration,} & \text{medium} \\ & \text{sales volume,} & \text{low} \end{bmatrix} = \begin{bmatrix} D, & c_6, & \text{medium} \\ & c_7, & \text{low} \end{bmatrix}
$$

where M_{im} is the nonmaterial part and M_{re} is the material part.

According to the particular domain knowledge, the following correlative network can be obtained:

$$\left.\begin{pmatrix} D, & c_3, & v_3 \\ D, & c_4, & v_4 \\ D, & c_5, & v_5 \\ D, & c_6, & v_6 \end{pmatrix}\right\} \sim \left(D, \quad c_1, \quad v_1 \right) \sim \left(D, \quad c_2, \quad v_2 \right)$$

Suppose that, according to the industrial experience of the enterprise, it is found that the main factors affecting the sales of this particular model of mobile phones are cost effectiveness and public praise:

$$\left.\begin{pmatrix} D, & c_1, & v_1 \\ D, & c_2, & v_2 \end{pmatrix}\right\} \sim \left(D, \quad c_7, \quad v_7 \right).$$

According to the material and nonmaterial transformation rule, a transformation of the matter-element of the nonmaterial part or the material part might lead to one conductive transformation in its matter-element, whatever the nonmaterial part or the material part is. That is, if active transformations $\varphi = \varphi_3 \wedge \varphi_4 \wedge \varphi_5 \wedge \varphi_6$ are conducted on the two parts of matter-elements:

$$\varphi_3 \left(D, \quad c_3, \quad v_3 \right) = \left(D, \quad c_3, \quad v_3' \right)$$

$$\varphi_4 \left(D, \quad c_4, \quad v_4 \right) = \left(D, \quad c_4, \quad v_4' \right)$$

$$\varphi_5 \left(D, \quad c_5, \quad v_5 \right) = \left(D, \quad c_5, \quad v_5' \right)$$

$$\varphi_6 \left(D, \quad c_6, \quad v_6 \right) = \left(D, \quad c_6, \quad v_6' \right),$$

then there must have the following conductive transformations: $\varphi \Rightarrow {}_\varphi T_1 \Rightarrow {}_1 T_2$.
Now, apply

$$_\varphi T_1 \left(D, \quad c_1, \quad v_1 \right) = \left(D, \quad c_1, \quad v_1' \right)$$

$$_1 T_2 \left(D, \quad c_2, \quad v_2 \right) = \left(D, \quad c_2, \quad v_2' \right),$$

so we have ${}_\varphi T_1 \wedge {}_1 T_2 \Rightarrow T$, applying $T (D, \quad c_7, \quad v_7) = (D, \quad c_7, \quad v_7')$, and $v_7' > v_7$.
Thus, it can be seen that the mobile phone's cost performance can be improved through changing the values of the price, system performance, software innovation, or the level of hardware configuration to improve public praise, ultimately resulting in increasing sales volume of the mobile phone.

Example 5.4.8

A lighting company expects to create a new product by changing the connection of the lamp components in the course of innovating a wall lamp D. How could the company carry out the planned creation?

The following analysis is conducted with the soft and hard conjugate transformation method. First, let us analyze the hard part of the wall lamp D. The wall lamp D consists of the lamp holder D_1, the lamp bulb D_2, and the lamp shade D_3, all of which can be expressed as matter-elements as:

$$
M_{hr} = \begin{bmatrix} D_1, & \text{shape,} & \text{L} \\ & \text{color,} & \text{white} \\ & \text{material,} & \text{stainless steel} \\ & \text{effect,} & \text{brace} \\ & \text{weight,} & \text{200 g} \end{bmatrix} \wedge \begin{bmatrix} D_2, & \text{shape,} & \text{sphere} \\ & \text{colors of light,} & \text{yellow} \\ & \text{material,} & \text{glass} \\ & \text{effect,} & \text{lighting} \\ & \text{weight,} & \text{50 g} \end{bmatrix}
$$

$$
\wedge \begin{bmatrix} D_3, & \text{shape,} & \text{cylinder} \\ & \text{color,} & \text{white} \\ & \text{material,} & \text{plastic} \\ & \text{effect,} & \text{spotlight} \\ & \text{weight,} & \text{100 g} \end{bmatrix} = M_{hr1} \wedge M_{hr2} \wedge M_{hr3}.
$$

The soft part of the wall lamp D includes various relations between the lamp holder D_1, the lamp bulb D_2, and the lamp shade D_3. This can be expressed using relation-elements as:

$$
M_{sf} = \begin{bmatrix} \text{Up-down position relation,} & \text{antecedent,} & D_2 \\ & \text{consequent,} & D_1 \end{bmatrix} \wedge \begin{bmatrix} \text{Helical relation,} & \text{antecedent,} & D_2 \\ & \text{consequent,} & D_1 \\ & \text{degree,} & \text{tightly} \end{bmatrix}
$$

$$
\wedge \begin{bmatrix} \text{Up-down position relation,} & \text{antecedent,} & D_3 \\ & \text{consequent,} & D_1 \end{bmatrix} \wedge \begin{bmatrix} \text{Intercalation relation,} & \text{antecedent,} & D_3 \\ & \text{consequent,} & D_1 \\ & \text{degree,} & \text{tightly} \end{bmatrix}
$$

$$
\wedge \begin{bmatrix} \text{Up-down position relation,} & \text{antecedent,} & D_3 \\ & \text{consequent,} & D_2 \end{bmatrix}
$$

$$
= M_{sf1} \wedge M_{sf2} \wedge M_{sf3} \wedge M_{sf4} \wedge M_{sf5}.
$$

By using the soft and hard conjugate analysis, it can be readily seen that the creation of new products is realizable through one transformation of the matter-elements of the hard part or relation-elements of the soft part. For example, the following active transformation is implemented on M_{hr1}:

$$\varphi M_{hr1} = \varphi \begin{bmatrix} D_1, & \text{shape}, & \text{L} \\ & \text{color}, & \text{white} \\ & \text{material}, & \text{stainless steel} \\ & \text{effect}, & \text{brace} \\ & \text{weight}, & \text{200 g} \end{bmatrix}$$

$$= \begin{bmatrix} D_1', & \text{shape}, & \text{spherical} \\ & \text{color}, & \text{white} \\ & \text{material}, & \text{stainless steel} \\ & \text{effect}, & \text{brace} \\ & \text{weight}, & \text{50 g} \end{bmatrix} = M_{hr1}'.$$

According to the correlation rule, this transformation inevitably leads to the conductive transformation of other relevant components (the hard part) and relations (the soft part). That is, there must be a conjugate transformation, such as

$$\varphi T_1 M_{sf} = \begin{bmatrix} \text{Front-back relation}, & \text{antecedent}, & D_2 \\ & \text{consequent}, & D_1' \end{bmatrix} \wedge \begin{bmatrix} \text{Helical relation}, & \text{antecedent}, & D_2 \\ & \text{consequent}, & D_1' \\ & \text{degree}, & \text{tightly} \end{bmatrix}$$

$$\wedge \begin{bmatrix} \text{Front-back relation}, & \text{antecedent}, & D_3 \\ & \text{consequent}, & D_1' \end{bmatrix} \wedge \begin{bmatrix} \text{Intercalation relation}, & \text{antecedent}, & D_3 \\ & \text{consequent}, & D_1' \\ & \text{degree}, & \text{tightly} \end{bmatrix}$$

$$\wedge \begin{bmatrix} \text{Front-back relation}, & \text{antecedent}, & D_3 \\ & \text{consequent}, & D_2 \end{bmatrix}$$

$$= M_{sf1}' \wedge M_{sf2}' \wedge M_{sf3}' \wedge M_{sf4}' \wedge M_{sf5}' = M_{sf}'.$$

The soft and hard conjugate transformation shows that changing the value of the shape of the hard part may result in one conductive transformation of the relation in soft parts to get a creative idea for new products.

Example 5.4.9

To avoid the fierce competition in an old product market and to open a new market, a certain pan-producing enterprise decided to improve the enterprise's competitiveness by researching a non-stick pan that meets people's demands with a material that is harmless to health. Please analyze the enterprise by using the latent and apparent conjugate pair method.

Let the new product concept developed by enterprise E be $O(t)$. According to the latent and apparent conjugate analysis, let $O(t_0)$ be the latent matter of the enterprise's product before the enterprise's production (i.e. t_0 moment). The product's latent matter-element before production (i.e. the matter-element in the t_0 moment) is denoted as:

$$
M_{lt}(O(t_0)) = \begin{bmatrix}
O(t_0), & \text{name,} & \text{bionic non-stick pan} \\
& \text{characteristic,} & \text{no oil fume} \wedge \text{non-stick pan} \\
& \text{material,} & \text{synthetic materials with ceramic and steel}
\end{bmatrix}.
$$

After identifying the new product and applying the patent, a market analysis and feasibility study indicates that the bionic non-stick pan will have a large market. The enterprise will take measures to make the above matter-elements apparent in the latent part, i.e. turning the latent part into the matter-element in time moment t_1: $\varphi M_{lt}(O(t_0)) = M_{ap}(O(t_1))$.

The above transformation φ causes matter-elements of the latent part to be transformed into the following matter-elements of the apparent part:

$$
M_{ap}(O(t_1)) = \begin{bmatrix}
O(t_1), & \text{name,} & \text{bionic non-stick pan} \\
& \text{characteristic,} & \text{no oil fume} \wedge \text{non-stick pan} \\
& \text{material,} & \text{materials compounded whit ceramic and steel} \\
& \text{patent number,} & a
\end{bmatrix}.
$$

That is, the planned bionic non-stick pans are produced.

In addition, suppose the new product must be accompanied by a new production workshop or a new production line. That is, the implementation of φ inevitably results in enterprise E establishing a new production line. This addition of a production line requires increased investment input by the enterprise. In addition, the enterprise can only profit from the product after the new product reaches the market. In other words, a series of conductive transformations will be generated because of the occurrence of the latent and apparent transformation. Only when profits are earned with the new product can the enterprise prove that the latent and apparent conjugate transformation is successful.

Example 5.4.10

Waste gas O_1, waste water O_2, and waste residue O_3 are produced by enterprise E in its production process. These are the negative parts of the enterprise E in terms of the enterprise's profit (marked as characteristic c), forming three negative part matter-elements.

$$
M_{1\text{ng}_c} = \begin{bmatrix}
O_1, & \text{major components,} & \text{gas} \\
 & \text{form,} & \text{gaseous} \\
 & \text{flammability,} & \text{good} \\
 & \text{color,} & \text{black} \\
 & \text{application,} & \text{none}
\end{bmatrix}
$$

$$
M_{2\text{ng}_c} = \begin{bmatrix}
O_2, & \text{major components,} & \text{harmful substance} \\
 & \text{form,} & \text{liquid} \\
 & \text{color,} & \text{brownish black} \\
 & \text{separability,} & \text{good} \\
 & \text{application,} & \text{none}
\end{bmatrix}
$$

$$
M_{3\text{ng}_c} = \begin{bmatrix}
O_3, & \text{major components,} & \text{SiO}_2 \\
 & \text{form,} & \text{solid} \\
 & \text{color,} & \text{tattletale gray} \\
 & \text{application,} & \text{none}
\end{bmatrix}
$$

These three wastes have become a heavy burden on the enterprise. To solve these problems, according to the characteristics of these three negative matter-elements, they are transformed respectively.

$\varphi_1 O_1 = \{O_1', O_1''\}$, i.e. to collect O_1 for separating the fuel gas O_1';

$\varphi_2 O_2 = \{O_2', O_2''\}$, i.e. to treat O_2 in a closed circuit without discharging outside to filter out industrial water O_2';

$\varphi_3 O_3 = O_3 \oplus \text{binder} \oplus \text{ingredients} = O_3'$, i.e. to add the binder and other ingredients into O_3 to form a new material O_3'.

Then, the following three positive matter-elements are formed:

$$
\varphi_1 T_1 M_{1\text{ng}_c} = M_{1\text{ps}_c} = \begin{bmatrix}
O_1', & \text{major components,} & \text{gas} \\
 & \text{form,} & \text{gaseous} \\
 & \text{flammability,} & \text{good} \\
 & \text{color,} & \text{none} \\
 & \text{application,} & \text{firing boiler} \wedge \text{generating electricity}
\end{bmatrix}
$$

$$\varphi_2 T_2 M_{2ng_e} = M_{2ps_e} = \begin{bmatrix} O_2', & \text{major components,} & \text{harmless water} \\ & \text{form,} & \text{liquid} \\ & \text{color,} & \text{none} \\ & \text{application,} & \text{production} \end{bmatrix}$$

$$\varphi_3 T_3 M_{3ng_e} = M_{3ps_e} = \begin{bmatrix} O_3', & \text{major components,} & SiO_2 \\ & \text{form,} & \text{solid} \\ & \text{color,} & \text{tattletale gray} \\ & \text{application,} & \text{use for producting bricks} \end{bmatrix}$$

Through these three transformations, the three wastes are turned into three treasures, significantly reducing costs and turning big profits for the enterprise. In other words, the negative part of the enterprise is transformed into a positive part. The specific approaches used are:

1. Collect the waste gas for separation and utilize it as fuel gas for boilers and power generation. Doing so reduces the fuel costs for the enterprise.
2. A fully closed-loop waste-water recycling system is established to produce clean, harmless water for reuse as industrial water, reducing the water consumption of the enterprise.
3. Waste residue can be used as raw materials of bricks, reducing the cost of buying bricks to expand the plant. Residue bricks can also be sold for earning profits for the enterprise.

5.5 Composite Transformation Methods

QUESTIONS AND THINKING:

■ How can a normal person move a heavy safe weighing 100 kg from his living room to his bedroom without scratching the floor?
■ Is it possible to adopt the transformation methods mentioned previously to solve this problem?
■ How many kinds of transformation methods do you have?

Complex computing and conductive transformations are often needed in the process of solving practical problems, known as composite transformation methods. The concepts of intermediating transformation and loss-compensating transformation will be briefly introduced.

5.5.1 *Intermediating Transformation Method*

Intermediating transformation refers to using a basic-element as a mediating effect when a certain transformation cannot accomplish the desired goal to transform the goal through the transformation. Each intermediating transformation is a special kind of PRODUCT transformation.

Generally speaking, assume that a basic-element B_0 is given. If transformation T is implemented to prevent $TB_0 = B$ from realization, φ can be introduced to make $\varphi B_0 = B_1$, which makes $T_1 B_1 = B_2$ and $T_2 B_2 = B$, and thus we have

$$(T_2 T_1 \varphi) B_0 = T_2 T_1 (\varphi B_0) = T_2 (T_1 B_1) = T_2 B_2 = B.$$

The transformation φ is called an intermediating transformation, while the basic-element B_1 is called an intermediating basic-element. Note that, if $T_1 B_1 = B$, it is unnecessary to transform T_2 any more.

For example, Americans landed on the moon by using the Apollo manned spacecraft in July 1969. The manned spacecraft is "an intermediate matter." Both "bridge" and "ship" are "intermediates" for realizing the goal of facilitating the crossing of a great river. It is an intermediating matter-element when it is represented by the matter-element.

Intermediating transformations have been applied in product innovations to obtain creative ideas for new products. Many "intermediating products" have been produced when using this method.

Case Analysis:

Example 5.5.1

The original practice for transferring pictures and files from a mobile phone to a computer involved connecting the phone to the computer with a specifically designed cable. In this situation, the connecting cable is an intermediate. Using this method requires carrying the connecting cable all the time. This is not convenient. Is there any other way to resolve this issue?

To solve this contradictory problem, the company Tencent added a function called "File Assistant" in QQ, its instant-messaging platform. Through the File Assistant, files in a phone can be directly transferred to a computer via a wireless network connection, and vice versa. The File Assistant software is an intermediate.

Suppose $M_1 = $ (File D, position, mobile phone), if we want to make the following transformation realize

$$T_1 M_1 = \left(\text{File } D, \quad \text{position}, \quad \text{mobile phone}\right) = \left(\text{File } D, \quad \text{position}, \quad \text{computer}\right) = M_1'.$$

However, this goal cannot be realized directly. We can first make a transformation:

$$\varphi M_1 = \varphi\left(\text{File } D, \quad \text{position}, \quad \text{mobile phone}\right)$$

$$= \left(\text{File } D, \quad \text{position}, \quad \text{"File Assistant" in QQ}\right) = M_0,$$

and then we can use the following transformation:

$$T_2 M_0 = T_2 \left(\text{File } D, \quad \text{position}, \quad \text{"File Assistant" in QQ} \right)$$

$$= \left(\text{File } D, \quad \text{position}, \quad \text{computer} \right) = M_1'.$$

Thus, transformation T_1 can be realized, i.e. $T_1 = T_2 \varphi$, where φ is the intermediating transformation, and the following is an intermediating matter-element:

$$M_0 = \left(\text{File } D, \quad \text{position}, \quad \text{"File Assistant" in QQ} \right).$$

From the perspective of conjugate analysis, the connecting cable in the above example is a material intermediate, while File Assistant is a nonmaterial intermediate. This shows that the creation of intermediates for solving contradictory problems can consider not only the material part, but also the nonmaterial part.

As another example, the sharing software for a printer is also an intermediate that allows more than one computer to share one printer. USB disks and removable hard disks are intermediates for realizing file transfers from different memory devices.

5.5.2 Loss Compensating Transformation Method

In the process of dealing with contradictions or product innovation, we often use the method of making up for deficiencies with redundance, which is called the method of loss-compensation transformation. "Exchange of goods and exchange of need goods" are matters' loss compensations. A wolf is good at running due to its long legs, but it's not very clever, whereas a wolf-like animal with short forelegs may be weak in running due to its short legs, but it may be clever and thoughtful. The wolf and the wolf-like animal utilize each other's strength to do bad things together, which is the origin of the idiom "acting in collusion with each other."

In the following we discuss two forms of loss-compensating transformation by taking a matter-element as an example.

5.5.2.1 Loss-compensating Transformation Among Matter-elements of the Same Characteristic

Given two matter-elements $M_1 = (O_1, \quad c, \quad v_1)$ and $M_2 = (O_2, \quad c, \quad v_2)$, if

$$T_1 M_1 = \left\{ M_{11}, \quad M_{12} \right\} = \left\{ \left(O_1', \quad c, \quad v_{11} \right), \left(O_2'', \quad c, \quad v_{12} \right) \right\}, \text{ and}$$

$$T_2 M_2 = M_2 \oplus M_{11} = \left(O_2, \quad c, \quad v_2 \right) \oplus \left(O_1', \quad c, \quad v_{11} \right) = \left(O, \quad c, \quad v \right) = M,$$

$T = T_2 T_1$ is called a loss compensation among matter-elements with the same characteristic.

The essence of a loss-compensating transformation is to decompose a part from one matter to supplement another matter. The general adding transformation can also be regarded as a special loss-compensating transformation. That is, a matter is directly supplemented to another matter.

When a loss-compensation transformation is conducted, one conductive transformation must occur in the value about the corresponding characteristic, i.e. one conductive transformation must occur among elements of the basic-element.

5.5.2.2 Loss-compensating Transformation Among Matter-elements with Different Characteristics

Given matter-elements

$$
M_1 = \begin{bmatrix} O_1, & c_1, & v_{11} \\ & c_2, & v_{12} \end{bmatrix} \quad \text{and} \quad M_2 = \begin{bmatrix} O_2, & c_1, & v_{21} \\ & c_2, & v_{22} \end{bmatrix},
$$

if the following transformation are implemented

$$
T_{11}\big(O_1, \quad c_1, \quad v_{11}\big) = \big(O_1', \quad c_1, \quad v_{11} \ominus v_{11}'\big),
$$

$$
T_{21}\big(O_2, \quad c_1, \quad v_{21}\big) = \big(O_2', \quad c_1, \quad v_{21} \oplus v_{11}'\big),
$$

$$
T_{22}\big(O_2, \quad c_2, \quad v_{22}\big) = \big(O_2, \quad c_2, \quad v_{22} \ominus v_{22}'\big),
$$

$$
T_{12}\big(O_1, \quad c_2, \quad v_{12}\big) = \big(O_1', \quad c_2, \quad v_{12} \oplus v_{22}'\big),
$$

M_1 and M_2 become

$$
M_1' = \begin{bmatrix} O_1', & c_1, & v_{11} \ominus v_{11}' \\ & c_2 & v_{21} \oplus v_{22}' \end{bmatrix}, \quad M_2' = \begin{bmatrix} O_2', & c_1, & v_{21} \oplus v_{11}' \\ & c_2 & v_{22} \ominus v_{22}' \end{bmatrix}.
$$

The transformation $T = (T_{21}T_{11}) \wedge (T_{12}T_{22})$ is a loss-compensating transformation among matter-elements of different characteristics.

Applying a loss-compensating transformation is essentially conducting a removing transformation for the value of a certain object on a certain characteristic. An adding transformation is conducted simultaneously on the value of another object on the characteristic. In many practical cases, one may conduct an adding transformation to realize loss compensation rather than conducting a removing transformation, or one may combine objects of two matter-elements into a new object.

In particular, for the same matter with values of two different characteristics, a removing transformation can be conducted on one of them while an adding transformation conducted on the other. This is called a loss-compensating transformation. There are many applications of the loss-compensating transformation method in the integration of enterprises.

Case Analysis:

Example 5.5.2

The hardness of an ordinary cutter cannot meet the hardness requirement for a processing job. The No. 45 steel cutter is therefore coated in an alloy, which can be formalized as:

$$M_1 = \left(\text{No. 45 steel cutter } D_1, \quad \text{hardness}, \quad 250\right),$$

$$M_2 = \left(\text{Hard alloy } D_2, \quad \text{hardness}, \quad 450\right).$$

Define

$$T_1 M_1 = M_1 \otimes M_2 = \left(\text{No. 45 steel cutter } D_1, \quad \text{hardness}, \quad 250\right) \otimes \left(\text{Hard alloy } D_2, \quad \text{hardness}, \quad 450\right)$$

$$= \left(\text{Hard alloy cutter } D, \quad \text{hardness}, \quad 450\right) = M.$$

Then T_1 is a loss-compensating transformation of the matter. Obviously, the corresponding conductive transformation is also conducted on the value of hardness.

Example 5.5.3

Chinese company D_1 cooperates with Japanese company D_2 to set up a factory in China. The Chinese side can utilize the advanced technology and equipment from the outside, while the foreign side can utilize the cheap labor capital and vast market of the Chinese side. The situation can be expressed with the following formalized model:

$$M_1 = \begin{bmatrix} D_1, & \text{level of technology,} & \text{low} \\ & \text{advancement of equipment,} & \text{backward} \\ & \text{cost of labor force,} & \text{cheap} \\ & \text{market conditions within country} & \text{vast} \\ & \text{location,} & \text{Chinese Mainland} \end{bmatrix}$$

$$M_2 = \begin{bmatrix} D_2, & \text{level of technology,} & \text{high} \\ & \text{advancement of equipment,} & \text{advanced} \\ & \text{cost of labor force,} & \text{costly} \\ & \text{market conditions within country} & \text{limited} \\ & \text{location,} & \text{Japan} \end{bmatrix}$$

The conducting transformation

$$TM_1 = M_1 \otimes M_2 = \begin{bmatrix} D, & \text{level of technology,} & \text{high} \\ & \text{advancement of equipment,} & \text{advanced} \\ & \text{cost of labor force,} & \text{low} \\ & \text{market conditions within country} & \text{vast} \\ & \text{location,} & \text{Chinese Mainland} \end{bmatrix} = M$$

shows that a Sino-Japanese joint venture D is established in mainland China by taking advantage of the high technical level and advanced equipment of the Japanese company and by taking advantage of the vast Chinese market and the low labor cost of the Chinese company.

From this example, it can be seen that there are many extension transformations for innovating or solving contradictory problems. In addition to basic extension transformations, there are operations of transformations and composite methods. It is important to analyze problems case by case and to select an appropriate transformation method to generate a strategy for solving contradictory problems based on different goals and conditions of problems.

Thinking and Exercises

1. What are the functions of a big tree? Is it possible to explore additional functions by utilizing the extensible analysis and the basic extension transformation method, such as charging, shading, lighting, and photosynthesis?
2. Wooden buckets can be used for filling water, rice, sand, etc. What else can be done with the bucket? Is it possible to use the extensible analysis and basic extension transformations to explore additional functions for the bucket?
3. If you have a small villa with a small window, the window cannot meet the needs of indoor lighting. What can you do to remedy the situation? Can you adjust the lighting of the house according to the change of weather, season, or even the mood of the inhabitants with random transformation and free control?

4. In the picture below, the baby would like to get off the bed by himself, but the bed is too high for him to get down safely. Please help the baby develop several ideas for solving this problem.

5. Can living plants be illuminated? Use methods of extensible analysis and extension transformations to obtain one or more ideas, and explain what kind of expansion analysis methods and extension transformations you used.

6. Can we eat a pen? Can a pen be do-it-yourself (DIY)? How can you obtain one or more creative ideas?

7. Can a mobile phone be DIY freely? How can you obtain one or more creative ideas?

8. Can you make four-page color document into a JPG file smaller than 200 kb, with the requirement that the text on the pages remains clear enough to be read?

9. Through the analysis of existing laser pointers, as given in this chapter, use extension transformation methods to generate creative ideas for new products.

10. Use the conjugate pair method to analyze your conjugate parts and implement the conjugate transformation on the shortcomings of your certain conjugate parts. Then analyze the conjugate transformation you implemented and describe solutions to make change(s).

11. Conduct transformations on the disadvantage parts of one of your resources with a conjugate transformation to find methods to make changes to your resources.

12. Represent the Monkey King's "72 changes" by using the method of classification of extension transformations.

Chapter 6

Evaluation and Selection of Creative Ideas
Superiority Evaluation Method (SEM)

Content Summary

The superiority evaluation method (SEM) is a basic method to quantitatively evaluate an object, including matter, affair, transformation, innovative idea, scheme, and so on. In this chapter, we first briefly introduce some essential preliminary knowledge. Then we successively present single level–single evaluating characteristic superiority evaluation, single level–multiple evaluating characteristic superiority evaluation, and multiple level–multiple evaluating characteristic superiority evaluation.

6.1 Pre-knowledge

QUESTIONS AND THINKING:

- You need to travel from Beijing to Guangzhou as quickly as possible, but all of the direct flights have been sold out. What should you do? How many methods are available for you to choose from? Which scheme is better? How do you select a measurable indicator? Is there any quantitative tool?
- How do you select an appropriate idea from a set of multiple innovative ideas for product innovation?
- Many contradictory problems have multiple solutions. How do you evaluate them quantitatively? Which solution is better?

6.1.1 Measurable Indicator

To evaluate an object, having a measurable indicator—that is, standards by which to evaluate the object—is essential. Generally, any given object can be measured as beneficial by using some measurable indicators, and it can deemed to be harmful by using other measurable indicators.

Therefore, evaluating an object requires quantitative degrees of the object's advantage or disadvantage and their possible variations. We should set up scientific evaluation standards that satisfy technological, economic, and social requirements according to our actual demands, then determine a measurable indicator $MI = \{MI_1, MI_2, ..., MI_n\}$, where $MI_i = (c_i, V_i)$ is a characteristic-element, c_i is an evaluation characteristic, and V_i is a quantified measurement value field ($i = 1, 2, ..., n$).

The selection of measurable indicators is critical. The principles of this selection include:

1. The aim of evaluation: There are different measurable indicators for different evaluating objects and subjections.
2. The comprehensiveness of evaluation: Requirements of technology, economy and society should be considered.
3. The feasibility of evaluation: Measurable indicators should be representative, and corresponding data should be reliable.
4. The stability of evaluation: Selected measurable indicators should be relatively stable. We need to be cautious about the measurable indicators that are significantly influenced by accidental factors. In other words, we should select measurable indicators that must be satisfied; other measurable indicators are selectable.

Technical requirement: This mainly involves the degree of difficulty during the implementation, the degree of innovation, or the users' request about the product's function.

Economic requirement: This is the requested capital for the implementation, and it is requested in terms of profit, such as manpower, material resource, money, and time.

Social requirement: This is the effect experienced by the entire social environment, including market requirements (object's potential market value and development prospect), environmental requirements (e.g. light, sound, wave, magnetism, and entities that may have an effect on our living circumstances), security requirements (e.g. information, estate, personal security, and so on), law requirements, social feedback, etc.

For instance, "buy a house" and "rent a house" correspond to different measurable indicators. Enterprises of different types may also select different measurable indicators for the same innovative idea of the product being considered. For more details about the selection of measurable indicators, please refer to Sections 6.2–6.4.

For many practical problems, some measurable indicators must be satisfied. Such measurable indicators are usually used to select the evaluating objects firstly.

Objects satisfying these measurable indicators will be further evaluated by other measurable indicators.

6.1.2 Dependent Function and Dependent Degree

In classical mathematics, a choice function is adopted to describe whether an object has certain property, of which the value takes only one of the two numbers 0 and 1, respectively, to describe "yes" or "no."

In fuzzy mathematics, a membership function is used to describe the degree by which an object satisfies a certain property, of which the value is taken from the interval [0,1].

In Extenics, the concept of dependence functions is proposed to describe the degree by which an object possesses a certain property. Specific formulas of dependence functions in the field of study have been established. They can quantitatively and objectively describe not only the degree by which an object possesses a certain property, but also its quantitative and qualitative process. A dependence function takes its value from the interval $(-\infty, +\infty)$.

For an object Z, let us see how to establish the corresponding dependence function $k(z)$ about a measurable indicator MI that describes the degree by which Z satisfies the given requirement. We call the value of $k(z)$ as the dependence degree of Z about MI.

When establishing the dependence function, several commonly used fields include:

1. Standard positive field, also called the satisfactory interval: $X_0 = <a_0, b_0>$;
2. Transition positive field, also called the acceptable interval: $X_+ = X - X_0 = <a, a_0> \cup <b_0, b>$;
3. Positive field: $X = <a, b>$;
4. Transition negative field: $X_- = <c, a> \cup <b, d>$;
5. Integrated interval of positive field and transition negative field, also called the augmented field or augmented interval, denoted as $\hat{X} = <c, d>$;
6. Negative field: $\bar{X} = \Re - X$; and
7. Standard negative field: $\bar{\bar{X}} = \Re - \hat{X}$.

The above intervals $X_0, X, \hat{X}$ satisfy $\hat{X} \supseteq X \supseteq X_0$ (the endpoints of the three intervals may overlap). When they have no common endpoints, their relationship is illustrated by Figure 6.1.1.

Figure 6.1.1   Relationship of intervals $<a_0, b_0>$, $<a, b>$, and $<c, d>$.

Division of the intervals is based on the values of dependence degree. The corresponding relationships between the ranges of the intervals and the dependence degree are:

$$x \in X_0 =< a_0, b_0 >, k(x) \geq 1;$$

$$x \in X_+ =< a, a_0 > \cup < b_0, b >, 0 \leq k(x) \leq 1,$$

$$x \in X_- =< c, a > \cup < b, d >, -1 \leq k(x) \leq 0; \quad \text{and}$$

$$x \in \bar{\bar{X}} = (-\infty, c > \cup < d, +\infty), k(x) \geq -1.$$

There are three types of commonly used dependence functions: elementary dependence function, simple dependence function, and discrete dependence function.

6.1.2.1 Elementary Dependence Function

Set a point x_0 in the satisfactory interval X_0 as the optimum point, and establish the elementary dependence function as follows:

$$k(x) = \begin{cases} \dfrac{\rho(x, x_0, X)}{D(x, x_0, X_0, X)}, & D(x, x_0, X_0, X) \neq 0, x \in X \\[2ex] -\rho(x, x_0, X_0) + 1, & D(x, x_0, X_0, X) = 0, x \in X_0 \\[2ex] 0, & D(x, X_0, X) = 0, x \notin X_0, x \in X \\[2ex] \dfrac{\rho(x, x_0, X)}{D(x, x_0, X, \hat{X})}, & D(x, x_0, X, \hat{X}) \neq 0, x \in \Re - X \\[2ex] -\rho(x, x_0, \hat{X}) - 1, & D(x, x_0, X, \hat{X}) = 0, x \in \Re - X \end{cases}$$

where $\rho(x, x_0, X)$ is called extension distance, describing the location relationship between an arbitrarily chosen point x and the interval X about fixed point x_0 (this is different from the concept of distance in the classical mathematics). $D(x, x_0, X_0, X)$ is called a place value, describing the location relationship between the point x about x_0 and the nested interval composed of X_0 and X; and at point x_0, $D(x_0, x_0, X_0, X) \neq 0$. Likewise, $D(x, x_0, X, \hat{X})$ is used to describe the location relationship between the point x about x_0 and the nested interval composed of X and $\hat{X}$.

Specifically, at point x_0, $D(x_0, x_0, X_0, X) = 0$, and for each $x \in X_0$, we specified $k(x) = -\rho(x, x_0, X_0) + 1$; in other intervals, the dependence functions have the same formulas as those presented above.

Such a dependence function satisfies:

a. When $x \in X_0$, $k(x) \geq 1$;
b. When $x \in X - X_0$ and $x \neq a \vee b$, $0 < k(x) \leq 1$;
c. When $x = a \vee b$, $k(x) = 0$; when $x = a_0 \vee b_0$, $k(x) = 1$;
d. When $x \in X_- =< c,a) \cup (b,d >, -1 \leq k(x) < 0$;
e. When $x = c \vee d$, $k(x) = -1$;
f. When $x \in \bar{X} = (-\infty,c) \cup (d,+\infty), k(x) < -1$; and
g. When $x = x_0$, $k(x)$ reaches the maximum.

Now we provide the relevant computation methods for extension distances and place values of the dependence function.

1. **Extension distance** $\rho(x,x_0,X)$: According to different locations of the fixed point x_0 in the interval, the extension distance can be categorized into left extension distance, middle extension distance, and right extension distance.

   Left extension distance: If $x_0 \in \left(a, \dfrac{a+b}{2} \right)$, the left extension distance is

   $$\rho_l(x,x_0,X) = \begin{cases} a-x, & x \leq a \\ \dfrac{b-x_0}{a-x_0}(x-a), & x \in< a,x_0 >. \\ x-b, & x \geq x_0 \end{cases}$$

   Specifically, when $x_0 = a$, we have

   $$\rho_l(x,a,X) = \begin{cases} a-x, & x < a \\ a_z, & x = a \\ x-b, & x > a \end{cases}$$

   where

   $$a_z = \rho_l(a,a,X) = \begin{cases} 0, & a \notin X \\ a-b, & a \in X \end{cases}.$$

   Right extension distance: If $x_0 \in \left(\dfrac{a+b}{2}, b \right)$, the right extension distance is

   $$\rho_r(x,x_0,X) = \begin{cases} a-x, & x \leq x_0 \\ \dfrac{a-x_0}{b-x_0}(b-x), & x \in< x_0,b >. \\ x-b, & x \geq b \end{cases}$$

Specifically, when $x_0 = b$, we have

$$\rho_r(x,b,X) = \begin{cases} a - x, & x < b \\ b_z, & x = b \\ x - b, & x > b \end{cases}$$

where

$$b_z = \rho_r(b,b,X) = \begin{cases} 0, & b \notin X \\ a - b, & b \in X \end{cases}.$$

Middle extension distance: If $x_0 = \dfrac{a+b}{2}$, the middle extension distance is

$$\rho_m(x,x_0,X) = \left| x - \frac{a+b}{2} \right| - \frac{b-a}{2} = \begin{cases} a - x, & x \leq \dfrac{a+b}{2} \\ x - b, & x \geq \dfrac{a+b}{2} \end{cases}.$$

Likewise, extension distance $\rho(x,x_0,X_0)$ and $\rho(x,x_0,\hat{X})$ in the elementary dependence function can be computed. Details are not repeated here.

2. **Place value:** The place value of x regarding interval X_0 and interval X about x_0, is:

$$D(x,x_0,X_0,X) = \rho(x,x_0,X) - \rho(x,x_0,X_0).$$

The place value of x regarding interval X and interval $\hat{X}$ about x_0, is:

$$D(x,x_0,X,\hat{X}) = \rho(x,x_0,\hat{X}) - \rho(x,x_0,X).$$

We have $D(x,x_0,X_0,X) \leq 0$, $D(x,x_0,X,\hat{X}) \leq 0$, in which the user should judge the type of extension distance according to the specific problem of concern.

Specifically, when there is no common endpoint in the nested intervals, we have $D(x,x_0,X_0,X) < 0$ and $D(x,x_0,X,\hat{X}) < 0$.

6.1.2.2 Simple Dependence Function

The commonly used simple dependence function satisfies the following four conditions:

1. The positive field is a bounded interval $X = <a, b>$ and has the maximum $M \in (a, b)$

$$k(x) = \begin{cases} \dfrac{x-a}{M-a}, & x \le M \\ \dfrac{b-x}{b-M}, & x \ge M \end{cases}.$$

When $M = a$

$$k(x) = \begin{cases} \dfrac{x-a}{b-a}, & x < a \\ \dfrac{b-x}{b-a}, & x \ge a \end{cases}.$$

When $M = b$

$$k(x) = \begin{cases} \dfrac{x-a}{b-a}, & x \le b \\ \dfrac{b-x}{b-a}, & x > b \end{cases}.$$

2. The positive field is an infinite interval $X = <a, +\infty)$, and has a maximum $M \in (a, +\infty)$

$$k(x) = \begin{cases} \dfrac{x-a}{M-a}, & x \le M \\ \dfrac{1+|M|}{x+1-M+|M|}, & x \ge M \end{cases}.$$

When $M = a$

$$k(x) = \begin{cases} x-a, & x < a \\ \dfrac{1+|a|}{x+1-a+|a|}, & x \ge a \end{cases}.$$

If the function $k(x)$ has no maximum in $X = <a, +\infty)$, we set $k(x) = x - a$.

3. The positive field is an infinite interval $X = (-\infty, b >$ and has a maximum $M \in (-\infty, b)$

$$k(x) = \begin{cases} \dfrac{1+|M|}{1+M-x+|M|}, & x \le M \\[3mm] \dfrac{x-b}{M-b}, & x \ge M \end{cases}.$$

When $M = b$

$$k(x) = \begin{cases} \dfrac{1+|b|}{1+b-x+|b|}, & x \le b \\[3mm] b-x, & x > b \end{cases}.$$

If the function $k(x)$ has no maximum in $X = (-\infty, b >$, we set $k(x) = b - x$.

4. The positive field is an infinite interval $X = (-\infty, +\infty)$ with a maximum $M \in X$

$$k(x) = \begin{cases} \dfrac{1}{1+M-x}, & x \le M \\[3mm] \dfrac{1}{x+1-M}, & x \ge M \end{cases}.$$

If the function $k(x)$ has no maximum in $X = (-\infty, +\infty)$, we set $k(x) = e^x$ or $k(x) = e^{-x}$.

6.1.2.3 Discrete Dependence Function

In many practical application problems, the values of the research object about some characteristics are discrete. For example, the quality grade of a product can be categorized as excellent, good, moderate or middle, and bad; students' performance can be categorized as superior, qualified, or unqualified. These discrete values are nonnumeric. The quality grade of a product can also be described as 1, 2, 3, or 4, of which the discrete values are numeric. In fuzzy mathematics, they usually take numbers between 0 and 1 as the membership function of the research object about a certain characteristic.

In Extenics, the dependence function describes the degree of how the research object satisfies the requirement about some characteristic. The dependence function is required to take values in $(-\infty, +\infty)$. Therefore, a discrete dependence function should take its values according to the specific practical problem.

For instance, a company's recruitment requires the value of the characteristic "organizing ability" to be better than "good." "Middle" denotes the critical state. We assume the value range of candidates about the characteristic is {excellent, good, middle, ordinary, bad}. We can then establish the following dependence function:

$$k(\mathrm{x}) = \begin{cases} 2, & x = \text{excellent}, \\ 1, & x = \text{good}, \\ 0, & x = \text{middle}, \\ -1, & x = \text{ordinary}, \\ -2, & x = \text{bad}. \end{cases}$$

When $k(x) > 0$, we consider the candidate to be qualified in the characteristic "organizing ability"; when $k(x) < 0$, we consider the candidate to be unqualified in the characteristic "organizing ability"; when $k(x) = 0$, we consider the candidate to be in the critical state about the characteristic "organizing ability." In practice, the critical state is regarded to be qualified sometimes, unqualified sometimes.

As another example, suppose a company's recruitment process requires candidates to pass the College English Test (CET)-Level 4 (with a score of 425 or higher). Obviously, "score 425" is the critical condition. Usually, $x = 425$ is regarded to be qualified. The corresponding dependent function is thus established as:

$$k(x) = \begin{cases} 1, & x > 425 \\ 0, & x = 425. \\ -1, & x < 425 \end{cases}$$

When $k(x) \geq 0$, the candidate x is regarded as qualified about the characteristic "English level." If necessary, the value of the characteristic can be more detailed, such as adding "passing CET-Level 6." Interested readers can establish the corresponding dependence functions.

Generally, a discrete dependent function takes the following form:

$$k(x) = \begin{cases} A_1(>0), & x = a_1 \\ A_2(>0), & x = a_2 \\ \cdots \\ A_k(>0), & x = a_k \\ 0, & x = a_0 \\ B_1(<0), & x = b_1 \\ B_2(<0), & x = b_2 \\ \cdots \\ B_l(<0), & x = b_l \end{cases}.$$

6.1.3 Scaled Dependence Degree

Assume that the dependence degree of the evaluating object $Z_j(j = 1,2,3,\ldots,m)$ about a measurable indicator $MI_i(i = 1,2,3,\ldots,n)$ is $k_i(x_j)$. Then we have

$$K_i(x_j) = \frac{k_i(x_j)}{\max\limits_{j \in \{1,2,\ldots,m\}} |k_i(x_j)|},$$

which is called the scaled dependence degree of Z_j about MI_i.

6.1.4 Superiority

For an evaluating object $Z_j(j = 1,2,3,\ldots,m)$, let the set of all measurable indicators except those measurable indicators that must be satisfied is $MI = \{MI_1, MI_2, \ldots, MI_n\}$, the dependence degree of Z_j about MI_i is $K_i(x_j)$ ($i = 1,2,3,\ldots,n$; $j = 1,2,3,\ldots,m$); and the weight coefficient of MI_i is α_i (α_i denote the relative importance of measurable indicator MI_i) ($i = 1, 2,\ldots, n$). Then we have:

$$\sum_{i=1}^{n} \alpha_i = 1, 0 \leq \alpha_i \leq 1.$$

For the following different requirements, we have different superiorities:

1. In the given practical problem, Z_j is regarded to be qualified only if the comprehensive dependence degree of all measurable indicators is larger than 0; the superiority is defined as

$$C(Z_j) = \sum_{i=1}^{n} \alpha_i K_i(x_j);$$

2. In the given practical problem, Z_j is regarded to be qualified as long as the dependence degree of a measurable indicator is larger than 0; then the superiority is defined as:

$$C(Z_j) = \bigvee_{i=1}^{n} K_i(x_j);$$

3. In the given practical problem, Z_j is regarded to be qualified only if the dependence degree of each measurable indicator is larger than 0; then the superiority is defined as

$$C(Z_j) = \bigwedge_{i=1}^{n} K_i(x_j);$$

4. In the given practical problem, if the dependence degree of a measurable indicator is strictly required to be larger than λ ($\lambda > 0$), the measurable indicator is called "a measurable indicator that must be satisfied." At such a time, we use the measurable indicator to carry out the initial round of evaluation. For all objects satisfying this measurable indicator, any of above three superiorities can be adopted.

6.2 SEM for a Single Measurable Indicator

The result of superiority evaluation stands for a relative concept, which is used to judge different objects' qualities about a same characteristic. The single-measurable indicator superiority evaluation method (SEM) usually selects one measurable indicator and evaluates the objects under consideration. The specific procedure is given as follows:

1. **Determine the measurable indicator:** Select a measurable indicator *MI* for the objects. The value range required by the measurable indicator is not always the satisfactory range, but it must be the acceptable range. Sometimes the satisfactory range and acceptable range can overlap.

2. **Initial evaluation:** Do some qualitative analysis on the measurable indicator, and judge whether it is a measurement that must be satisfied. If so, filter out the objects not satisfying this measurement and finish the evaluation; otherwise, go to the next step.

3. **Compute the dependence degrees as the superiorities of the objects to be evaluated:** Assume the selected measurable indicator is $MI = (c, X)$, where X denotes the acceptable range (this can be described by an acceptable interval). Establish the dependence function $K(x)$ according to the requirements of the measurable indicator. Compute the dependence degrees as the superiorities of the objects to be evaluated.

Before establishing the dependence function, we should determine the value range of the measurable indicator and judge whether it should be extended to involve the range that is unacceptable originally but can change to be acceptable with the variation of some objective conditions or can shrink to obtain a more desirable range.

If the acceptable interval X is a definite interval or an infinite interval (any value in the interval is acceptable), the unique maximal dependence degree is at the optimum point, and it is unnecessary to further divide X, then the simple dependence function is used.

Assume X is a set of some discrete data and MI denotes the qualitative level of the product of concern, so $X = \{$level A, level B, level C$\}$. A unit of the product is regarded to be qualified only if its level achieves A (the value is 1); B corresponds to the critical value, and C is regarded to be unqualified. We then have

$$k(x) = \begin{cases} 1, & x = \text{level A} \\ 0, & x = \text{level B} \\ -1, & x = \text{level C} \end{cases}$$

The values of various levels can be obtained according to the experts' suggestion or historical data.

If X is further considered to be a nested interval constituted by X_0, X, and $\hat{X}$, namely $X_0 \subseteq X \subseteq \hat{X}$, where X_0 is the satisfactory interval, then the elementary dependence function introduced in section 6.1 can be applied.

Case Analysis:

Example 6.2.1

For a type of workpiece, the optimum diameter is $M = 30$. Establish a dependence function to describe the degree of workpiece that satisfies the requirement of diameter ($\Phi 30^{+0.01}_{-0.02}$), and compute its values at 30.1, 29.99, 30, and 29, respectively.

Solution: Assume $X = <29.98, 30.01>$ is the diameter range satisfying the requirement, namely the positive field. According to the simple dependence function from section 6.1.2, we have:

$$k(x) = \begin{cases} \dfrac{x - 29.98}{30 - 29.98}, & x \le 30 \\[2mm] \dfrac{30.01 - x}{30.01 - 30}, & x \ge 30 \end{cases}$$

$$= \begin{cases} 50(x - 29.98), & x \le 30 \\ 100(30.01 - x), & x \ge 30 \end{cases}$$

When $x = 30.1$, $k(30.1) = 100(30.01 - 30.1) = -9$;
When $x = 29.99$, $k(29.99) = 50(29.99 - 29.98) = 0.5$;
When $x = 30$, $k(30) = 100(30.01 - 30) = 50(30 - 29.98) = 1$; and
When $x = 29$, $k(29) = 50(29 - 29.98) = -49$.

Eample 6.2.2

An enterprise plans to purchase some three-dimensional printers. There are three candidate brands, of which the prices are respectively 29,000 yuan, 27,000 yuan, and 32,000 yuan. Assume the price of the enterprise's desirable desk-top three-dimensional printer is 25,000 yuan; the acceptable price range is from 15,000 yuan to 30,000 yuan; the satisfactory price range is from 20,000 yuan to 27,000 yuan. Considering not very high requirements of the printer's quality and accuracy and some promotion activities, the printers with prices from 10,000 yuan to 15 thousand yuan are also acceptable. Considering the future profit the printers may bring to the enterprise or other possible financial sources, the printers with prices from 30,000 yuan to 35,000 yuan are also acceptable. Compute the enterprise's satisfactory degree about the three types of printers' prices.

Solution: According to given information, the optimum and various intervals of the enterprise about the price of desk-top three-dimensional printer are
Optimum: $x_0 = 25$ thousand yuan; positive field: $X_0 = <a, b> = <15, 30>$ thousand yuan;
Satisfactory interval (standard positive field): $X_0 = <a_0, b_0> = <20, 27>$ thousand yuan;
Acceptable interval (transition positive field): $X_+ = <a, a_0> \cup <b_0, b> = <15, 20> \cup <27, 30>$ thousand yuan; and
Transition negative field: $X_- = <c, a> \cup <b, d> = <10, 15> \cup <30, 35>$ thousand yuan.
The combination of positive field and transition negative field is:

$$\hat{X} = <c, d> = <10, 35> \text{ thousand yuan}$$

Obviously, X_0, X and $\hat{X}$ have no common endpoints, so we can use the elementary dependence function in section 6.1:

$$k(x) = \begin{cases} \dfrac{\rho(x, x_0, X)}{D(x, x_0, X_0, X)}, & D(x, x_0, X_0, X) \neq 0, x \in X \\[3mm] \dfrac{\rho(x, x_0, X)}{D(x, x_0, X, \hat{X})}, & D(x, x_0, X, \hat{X}) \neq 0, x \in \mathfrak{R} - X \end{cases}$$

Next, we compute the dependence degrees of the three-dimensional printers (namely superiorities) of various brands:

When $x = 29$,

$$k(29) = \frac{\rho(29, 25, X)}{D(29, 25, X_0, X)}$$

$$= \frac{29 - 30}{\rho(29, 25, X) - \rho(29, 25, X_0)}$$

$$= \frac{29 - 30}{(29 - 30) - (29 - 27)} = \frac{1}{3} \approx 0.33$$

When $x = 27$,

$$k(27) = \frac{\rho(27, 25, X)}{D(27, 25, X_0, X)}$$

$$= \frac{15 - 27}{\rho(27, 25, X) - \rho(27, 25, X_0)} = 1.0$$

When $x = 32$,

$$k(32) = \frac{\rho(32, 25, X)}{D(32, 25, X, \hat{X})}$$

$$= \frac{32 - 30}{\rho(32, 25, \hat{X}) - \rho(32, 25, X)}$$

$$= \frac{32 - 30}{(32 - 35) - (32 - 30)} = -0.4$$

The enterprise's satisfactory degrees about the three brands of three-dimensional printers are 0.33, 2.0, and –0.4, respectively. The second brand clearly has the highest satisfactory degree.

In fact, such a single-measurable indicator evaluation can be implemented directly by considering the difference between the optimum and the evaluating objects' values, but doing so cannot provide the accurate satisfactory degree.

This example has verified the fact that the concept of dependence functions is suitable to measure the degree by which the evaluating objects conform to the requirements.

Example 6.2.3

In automobile production, with regard to cost, security, technical difficulty, functional requirement, and so on, we have specific requirements about the thickness of automobiles' steel board. For a certain brand, we have two types of automobiles (ES and RX), of which the average thickness of steel board is 0.77 mm and 1.04 mm, respectively. About the thickness characteristic, evaluate the superiorities of the two types of automobiles.

1. Determine a measurable indicator: According to the relevant professional knowledge, the acceptable range of steel board's thickness is $<0.7, 2.0>$ mm, so we have

$$MI_1 = (\text{thickness, acceptable thickness range}) = (\text{thickness}, <0.7, 2.0> \text{mm}).$$

2. Initial valuation: The production factory has no special requirement for automobiles' steel board, so the thickness is not a measurable indicator that must be satisfied.
3. Compute the dependent degree, namely superiority: Shrink the acceptable range to obtain the satisfactory interval $X_{01} = <0.8, 1.9>$ mm; the positive field is $X_1 = <0.7, 2.0>$ mm; considering the special thickness requirement or improvement of technology, we regard respectively $<0.65, 0.7>$ mm and $<2.0, 2.2>$ mm as the transition negative field, obtaining the augmented field $\hat{X}_1 = <0.65, 2.2>$ mm.

Assume the optimum thickness is $x_{01} = 1.8$ mm; let Z_1, Z_2 respectively denote the ES and RX types of automobiles; the average thicknesses of Z_1, Z_2 are respectively $x_{11} = 0.77$ mm, $x_{12} = 1.04$ mm. Use the elementary dependence function to obtain the dependence degrees of Z_1, Z_2 about MI_1:

$$k_1(x_{11}) = \frac{\rho(x_{11}, x_{01}, X_1)}{D(x_{11}, x_{01}, X_{01}, X_1)} = \frac{0.7 - x_{11}}{\rho(x_{11}, x_{01}, X_1) - \rho(x_{11}, x_{01}, X_{01})}$$

$$= \frac{0.7 - 0.77}{(0.7 - 0.77) - (0.8 - 0.77)} = 0.7$$

$$k_1(x_{12}) = \frac{\rho(x_{12}, x_{01}, X_1)}{D(x_{12}, x_{01}, X_{01}, X_1)} = \frac{0.7 - x_{12}}{\rho(x_{12}, x_{01}, X_1) - \rho(x_{12}, x_{01}, X_{01})}$$

$$= \frac{0.7 - 1.04}{(0.7 - 1.04) - (0.8 - 1.04)} = 3.4$$

Clearly, Z_2 has a better superiority, so we conclude that the RX type of automobile has a more suitable thickness of steel board.

6.3 SEM for One-Level Multiple Measurable Indicators

When a set of objects needs to be evaluated, multiple factors involving economy, technology, and society should be considered. That is, one has to deal with a multiple-indicator comprehensive evaluation.

When multiple measurable indicators are simultaneously considered and there is no difference of grades among them, we adopt the so-called one-level, multiple-indicator SEM, briefly called the SEM. The specific procedure of using this method follows.

Determine the Measurable Indicators

For the measurable indicators MI_i and their value ranges V_i, the following points should be noticed:

1. The selection should be based on social and economic phenomenon, and should be based on the evaluating objects-related value range and historical data.
2. The selection should include the changing tendency of social and economic phenomenon, and the estimated values of change should be considered when determining the value ranges.
3. The decision of value ranges should be adjustable and manageable. Thus, the standard data, such as programming values, planning values in the national societies (districts, sectors), and economic management, can be regarded as the boundaries of the value ranges.

Determine the Weight Coefficients

To evaluate objects $Z_j (j = 1, 2, ..., m)$, the importance of various measurable indicators $MI_1, MI_2, ..., MI_n$ is different from one indicator to another, which is described by the corresponding weight coefficients of these indicators. The measurable indicators that must be satisfied are denoted by Λ. For other measurable indicators, their weight coefficients take values from [0, 1], denoted by $\alpha = (\alpha_1, \alpha_2, ..., \alpha_n)$. If $\alpha_{i_0} = \Lambda$, we have

$$\sum_{\substack{i=1 \\ i \neq i_0}}^{n} \alpha_i = 1.$$

The weight coefficients are significant to the superiority. Different weight coefficients provide different conclusions, leading to changes in the evaluating objects' superiority order. If the weight coefficients are determined artificially, the subjectivity of the researcher may influence the reliability and authenticity of the evaluation. To determine the weight coefficients reasonably, hierarchical analysis and other methods can be used to determine the order of the relative importance of measurable indicators.

Initial Evaluation

After determining the weight coefficients, the measurable indicators that must be satisfied are used to filter the objects initially. For all the objects satisfying Λ, execute the following procedural steps (assume $Z_1, Z_2, \ldots, Z_m$ are the qualified objects under the measurable indicators that must be satisfied).

Establish a Dependence Function and Compute the Dependence Degree

Assume that the set of measurable indicators are $MI = \{MI_1, MI_2, \ldots, MI_n\}$, $MI_i = (c_i, V_i), (i = 1, 2, \ldots, n)$; the weight coefficients are $\alpha = (\alpha_1, \alpha_2, \ldots, \alpha_n)$.

According to the requirement of each measurable indicator, establish the following dependence functions $k_1(x_1), k_2(x_2), \ldots, k_n(x_n)$. The establishing method for each measurable indicator is the same as that for a single indicator, which can be carried out by following the same steps.

Compute the Scaled Dependent Degree

Assume that the value of evaluating object Z_j about measurable indicator MI_i is x_{ij} and the dependence degree is $k_i(x_{ij})$. The scaled dependence degree is computed by using:

$$K_i(x_{ij}) = \frac{k_i(x_{ij})}{\max\limits_{j=1}^{m} \left| k_i(x_{ij}) \right|}, \quad (i = 1, 2, \ldots, n; j = 1, 2, \ldots, m).$$

Compute Superiority

Multiple measurable indicator superiority can be computed by using the method given in section 6.1.4.

The first type of superiority is computed according to Table 6.3.1. The second and third types of superiority are computed according to Table 6.3.2. The value of the superiority indicates the quality of the object.

Note: When dealing with practical problems, some indicators must be satisfied; otherwise, an object is unqualified no matter how excellent it is about other indicators. For example, in architecture designs, particular selection of material and allocation of equipment must satisfy the safety requirements. Unqualified material, equipment, and schemes cannot be used.

When evaluating an object, we should consider not only the beneficial terms but also the harmful terms. For example, an enterprise produces a product; despite its economic benefits, the exhaust gas, a byproduct of the production, can pollute the environment seriously. In comparison, another product provides less economic

Table 6.3.1 Superiority Table of the First Type of Superiority

Measurable Indicator	Weight Coefficient	Dependent Degree				Scaled Dependent Degree		
		Z_1	Z_2	$\cdots$	Z_m	Z_1	$\cdots$	Z_m
MI_1	α_1	$k_1(x_{11})$	$k_1(x_{12})$	$\cdots$	$k_1(x_{1m})$	$K_1(x_{11})$	$\cdots$	$K_1(x_{1m})$
MI_2	α_2	$k_2(x_{21})$	$k_2(x_{22})$	$\cdots$	$k_2(x_{2m})$	$K_2(x_{21})$	$\cdots$	$K_2(x_{2m})$
$\cdots$	$\cdots$	$\cdots$	$\cdots$	$\cdots$	$\cdots$	$\cdots$	$\cdots$	$\cdots$
MI_n	α_n	$k_n(x_{n1})$	$k_n(x_{n2})$	$\cdots$	$k_n(x_{nm})$	$K_n(x_{n1})$	$\cdots$	$K_n(x_{nm})$
Superiority						$\sum_{i=1}^n \alpha_i K_i(x_{i1})$	$\cdots$	$\sum_{i=1}^n \alpha_i K_i(x_{im})$

Table. 6.3.2 Superiority Table of the Second and Third Types of Superiority

Measurable Indicator	Dependent Degree				Scaled Dependent Degree		
	Z_1	Z_2	...	Z_m	Z_1	...	Z_m
MI_1	$k_1(x_{11})$	$k_1(x_{12})$	...	$k_1(x_{1m})$	$K_1(x_{11})$	...	$K_1(x_{1m})$
MI_2	$k_2(x_{21})$	$k_2(x_{22})$	...	$k_2(x_{2m})$	$K_2(x_{21})$	...	$K_2(x_{2m})$
...	...	...	...	...	...	...	...
MI_n	$k_n(x_{n1})$	$k_n(x_{n2})$	...	$k_n(x_{nm})$	$K_n(x_{n1})$	...	$K_n(x_{nm})$
Minimum superiority					$\bigwedge\limits_{i=1}^{n} K_i(x_{i1})$	...	$\bigwedge\limits_{i=1}^{n} K_i(x_{im})$
Maximum superiority					$\bigvee\limits_{i=1}^{n} K_i(x_{i1})$	...	$\bigvee\limits_{i=1}^{n} K_i(x_{im})$

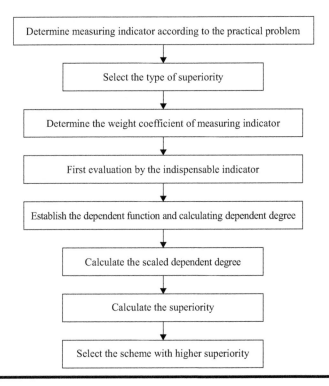

Figure 6.3.1 Basic flowchart of the one-level multiple indicator superiority evaluation method.

benefit, but it does not harm the environment. Thus, to decide which product to produce, we need to comprehensively consider producing a suitable scheme. Moreover, we should also consider the dynamic property and variable aspects of the product so that we can predict the potential advantages and disadvantages.

The measurable indicators for one-level superiority evaluation do not need to be divided into more detailed sub-indicators. Each measurable indicator is for a certain evaluating characteristic. A flowchart of one-level multiple measurable indicators superiority evaluation is illustrated in Figure 6.3.1.

Case Analysis:

Example 6.3.1

In example 6.2.3, we performed a superiority evaluation on ES and RX types of automobiles about the thickness of steel board. The conclusion was that the RX automobile is better. However, this result is based on only one evaluation characteristic. For a more comprehensive evaluation, other evaluation characteristics should be considered. Now we perform multiple indicator superiority evaluation on the two types of automobiles about the thickness of steel board, price, public appraisals, and so on.

1. **Determine measurable indicators:** In terms of technology, we regard the thickness of steel board as an evaluating characteristic. For details, refer to example 6.2.3.

   In terms of economy, we regard price as an evaluating characteristic. For people with moderate incomes, the acceptable range, namely the positive field, is $<0.1, 0.5>$ million yuan.

$$MI_2 = \left(\text{Price}, <0.1, 0.5> \text{ million yuan}\right)$$

   In terms of society, we regard the online feedback as an evaluation characteristic. Generally, the acceptable score, namely the positive field, is $<4.0, 5.0>$.

$$MI_3 = \left(\text{Evaluating score online}, <4.0, 5.0>\right)$$

2. **Determine the weight coefficients:** According to the importance of each of the three measurable indicators, the weight coefficients of MI_1, MI_2, MI_3 are determined to be $\alpha_1 = 0.3$, $\alpha_2 = 0.4$, $\alpha_3 = 0.3$, respectively.
3. **Initial evaluation:** Customers can propose special requirements. In this case, there are no special requirements. Therefore, this step is skipped.
4. **Compute dependence degrees and scaled dependence degrees:** The dependence degree of measurable indicator MI_1 has been given in example 6.2.3. The scaled dependence degrees of Z_1 and Z_2 are respectively given below:

$$K_1(x_{11}) = \frac{0.7}{3.4} = 0.206, \qquad K_1(x_{12}) = 1.$$

Compute the dependence degree and scaled dependence degree of measurable indicator MI_2.

   Because the positive field of price is $X_2 = <0.1, 0.5>$ million yuan, for the reasonability of evaluation, the value range can be further shrunk to obtain the satisfactory range, namely the standard positive field $X_{02} = <0.15, 0.3>$ million yuan. When the price is in $<0.07, 0.1>$ million yuan, it can be changed to be acceptable by some transformation, but if the price exceeds 0.5 million yuan, it is unacceptable. We can obtain the augmented interval $\hat{X}_2 = <0.07, 0.5>$ million yuan.

   Assume the optimum price of the automobiles with this brand for customers is $x_{02} = 0.3$ million yuan. The prices of Z_1, Z_2 are respectively $x_{21} = 0.359$ million yuan, and $x_{22} = 0.469$ million yuan. Using the formula of the elementary dependence function, we obtain the dependence degrees of Z_1, Z_2 about measurable indicator MI_2, respectively:

$$k_2(x_{21}) = \frac{\rho(x_{21}, x_{02}, X_2)}{D(x_{21}, x_{02}, X_{02}, X_2)} = \frac{x_{21} - 50}{\rho(x_{21}, x_{02}, X_2) - \rho(x_{21}, x_{02}, X_{02})}$$

$$= \frac{x_{21} - 50}{(x_{21} - 50) - (x_{21} - 30)} = 0.705$$

$$k_2(x_{22}) = \frac{\rho(x_{22}, x_{02}, X_2)}{D(x_{22}, x_{02}, X_{02}, X_2)} = \frac{x_{22} - 50}{\rho(x_{22}, x_{02}, X_2) - \rho(x_{22}, x_{02}, X_{02})}$$

$$= \frac{x_2 - 50}{(x_{22} - 50) - (x_{22} - 30)} = 0.155.$$

The scaled dependence degrees are

$$K_2(x_{21}) = \frac{0.705}{0.705} = 1, \quad K_2(x_{22}) = \frac{0.155}{0.705} = 0.22.$$

Compute dependence degree and scaled dependence degree about measurable indicator MI_3.

For the online feedback score, the acceptable range is $X_3 = <4.0, 5.0>$, the optimum is the full score 5; the scores of Z_1, Z_2 are respectively $x_{31} = 4.0$, $x_{32} = 4.0$; and the optimum is $x_{03} = 5.0$.

Using the formula of the simple dependence function, we can obtain the dependence degrees of Z_1, Z_2 about measurable indicator MI_3, respectively, as follows:

$$k_3(x_{31}) = \frac{x_{31} - 4}{x_{03} - 4} = 0, \quad k_3(x_{32}) = \frac{x_{32} - 4}{x_{03} - 4} = 0.$$

Therefore, the scaled dependence degrees are:

$$K_3(x_{31}) = 0, \quad K_3(x_{32}) = 0.$$

5. **Compute the superiority:** The comprehensive evaluation is implemented by the weighted sum of individual superiorities. The superiorities of Z_1, Z_2 respectively are

$$C(Z_1) = \sum_{i=1}^{3} \alpha_i K_i(x_{i1}) = 0.3 * 0.206 + 0.4 * 1 + 0 = 0.4618$$

$$C(Z_2) = \sum_{i=1}^{3} \alpha_i K_i(x_{i2}) = 0.3 * 1 + 0.4 * 0.22 + 0 = 0.388$$

From this we see that Z_1 has a better superiority, so we can conclude with the following result: When comprehensively considering the characteristics of thickness of steel board, price, and online evaluation, the ES automobile (Z_1) is better than the RX (Z_2).

6.4 SEM for Multilevel Measurable Indicators

When the evaluating object is complex, there are many measurable indicators involved. Each of them has its own weight. Therefore, when there are a lot of measurable indicators, the following problems may arise.

It can be difficult to allocate the weight coefficients. Because each allocation of weights usually depends on people's subjective judgment, when there are many measurable indicators, it is hard to obtain such a reasonable judgment. On the other hand, because the measurable indicators are hierarchical, it is hard to allocate the weights even if the hierarchical analysis method is adopted.

It can be hard to obtain meaningful results. Due to the normalization of the weight coefficients, when there are many measurable indicators, each weight coefficient becomes very small, making it hard to indicate its position among all the coefficients.

Thus, we adopt a method of multilevel superiority evaluation to resolve these problems. Each multilevel superiority evaluation is based on the superiority evaluation introduced above. First, the measurable indicators are classified into different grades. Then each grade is given its own weight. Finally, a comprehensive evaluation is performed.

6.4.1 Division of Multilevel Measurable Indicators and Decision of Weight Coefficients

A measurable indicator is used to judge the quality of objects and determines the correctness and reasonability of the entire SEM. Weight coefficients stand for a description of the importance of the evaluation characteristics for the evaluator, determine the relationship between the evaluation results and the evaluator's demand, and indicate the practicability of the SEM. Therefore, division of measurable indicators and decision of the weight coefficients influence the results of the evaluation directly. To ensure the correctness, reasonability, and practicability of the superiority evaluation, we introduce the methods of dividing multilevel measurable indicators and determining the weight coefficients.

Multilevel Measurable Indicators

Each multilevel measurable indicator has multiple evaluation characteristics. In a practical superiority evaluation, the diversity of evaluation objects and evaluators makes measurable indicators hierarchical and categorical. We should comprehensively consider multiple evaluation characteristics and establish a multilevel measurable indicator system when implementing a multilevel superiority evaluation.

When determining a multilevel measurable indicator, we should consider the aim, comprehensiveness, feasibility, and stability of the evaluation. An evaluation standard that satisfies the technical requirement, economic requirement, and social requirement should be made according to the given practical problem. Namely, we categorize measurable indicators and comprehensively consider the object's complex characteristics in terms of technology, economy, and society, while avoiding of possibility of leaving out the evaluation of some characteristics. For specific contents, see Section 6.1.1.

The requirements of these three aspects are not absolutely independent. The following problems should be noted when determining the relevant indicators:

1. The technical requirement, economic requirement, and social requirement are used to determine three different aspects of measurable indicators. Each aspect can involve multiple evaluation characteristics. Each evaluation characteristic can be further divided according to the given practical situation.

2. An evaluation characteristic may not simply belong to one aspect of the technical requirement, economic requirement, or social requirement. Instead, it could simultaneously belong to several aspects. In other words, the evaluation characteristic is generated by a combination of several aspects

3. A measurable indicator can be divided into sub-indicators. A sub-indicator can be further divided until it cannot not be further divided, thus forming a hierarchy of measurable indicators.

Combining three categories of measurable indicators, a multilevel measurable indicator system can be established as shown in Figure 6.4.1(a). For convenience and for making the structure of measurable indicators more clear, different levels of measurements can be defined as different grades of measurable indicators.

First-level indicators are directly determined by the three categories of requirements. Assume that

$$MI = \{MI_1, MI_2, \ldots, MI_n\}$$

are first-level indicators. Sub-indicators can be developed from the first-level indicators by division. Assume that

$$MI_{ij}(i = 1, 2, \ldots, n; j = 1, 2, \ldots, m_h; h = 1, 2, \ldots, r)$$

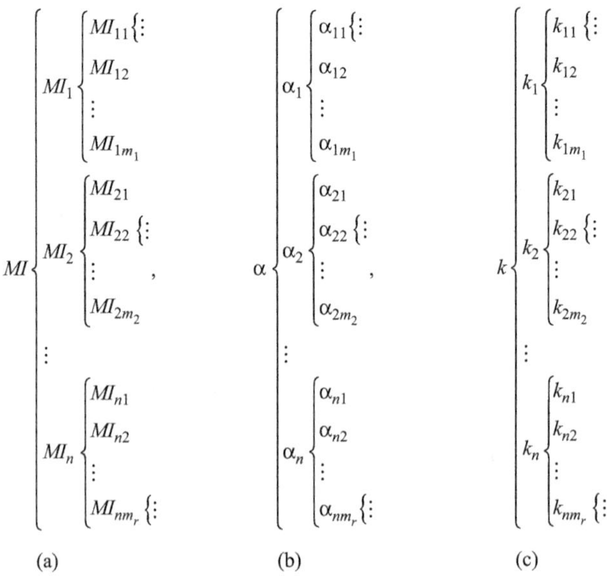

(a)   (b)   (c)

Figure 6.4.1   System structure of multilevel SEM.

are second-level indicators. The third- and fourth-level indicators can be constructed in the same manner. Sub-indicators originating from the same indicator are called homologous indicators. For example, $MI_{11}, MI_{12}, ..., MI_{1m}$ are originated from MI_1 and are homologous indicators.

Weight Coefficients

To evaluate an object, we should comprehensively consider the complex characteristics of the object. Different characteristics have different levels of importance to the evaluator. Different characteristics need to be assigned with different values according to their corresponding levels of importance. The division of importance into levels is performed among homologous indicators. That is, the division of the levels of importance of the evaluation characteristics is based on homologous indicators. Among homologous indicators, the levels of importance of various indicators are denoted by weight coefficients. The symbol of weight coefficient corresponds to that of the measurable indicator, as shown in Figure 6.4.1(b).

The decision of weight coefficients directly influences the result of the superiority evaluation. The methods for determining weight coefficients include the empirical weight method, the factor analysis weight method, the independence weight method, and the hierarchical analysis weight method. We should select a suitable method to determine the weight coefficients. To ensure the reasonableness of weight coefficients, the following principles should be satisfied when the weight coefficient of each measurable indicator is determined:

1. The method of determining the weight coefficients should be selected according to the evaluator's demand. A large weight coefficient indicates a high level of importance of the measurable indicator, and a small weight coefficient indicates a low level of importance.
2. Homologous weight coefficients correspond to homologous indicators. For example, $\alpha_{11}, \alpha_{12}, ..., \alpha_{1m_1}$ are the weights corresponding to the homologous indicators $MI_{11}, MI_{12}, ..., MI_{1m_1}$. Thus the sum of these homologous weight coefficients is 1.
3. The indicators that must be satisfied are denoted by Λ. For other measurable indicators, the weight coefficients are values from the interval [0,1] according to the levels of importance of the corresponding measurable indicators.

Once the weight coefficients are determined, the levels of importance of various characteristics are determined. The evaluator can change the weight coefficients according to his own demand to make the results of the evaluation more desirable.

6.4.2 Main Procedural Steps of the Multilevel SEM

The multilevel SEM evaluates the objects from the evaluator's perspective. It can help the evaluator make reasonable evaluations. The main procedural steps are explained here.

1. **Determine the measurable indicator system:** Using the dividing method introduced above, establish the measurable indicator system, shown in Figure 6.4.1(a).
2. **Determine the weight coefficients:** According to the principles of determining the weight coefficients, determine the weight coefficients of the measurable indicator system established in (1), as shown in Figure 6.4.1(b).
3. **Initial evaluation:** After determining the weight coefficients of various measurable indicators, first use the measurable indicators that must be satisfied to filter through the evaluating objects and delete the unsatisfied objects; then perform the following procedural steps on the remaining objects.
4. **Establish a dependence function for each measurable indicator and compute the dependence degree:** Establish dependence functions according to the specific situations, and compute the dependence degrees of each object about each measurable indicator.
5. **Compute scaled dependence degrees:** Different characteristics have different dimensions, leading to problem of dimensional difference and lack of uniformity. Therefore, normalization is essential. All levels of scaled dependence degrees are dimensionless dependence degrees. For a detailed method of computation, refer to Section 6.1.3. Each evaluating object's scaled dependence degree about each measurable indicator can be calculated.
6. **Compute the superiority:** Superiority $C(Z_j)$ is the comprehensive degree by which the evaluating object satisfies the requirement. Corresponding to a multilevel measurable indicator, a multilevel superiority also exists. The subscript of a multilevel superiority is the same as that of the corresponding measurable indicator.

Case Analysis

Multilevel SEM is widely applied to industrial production and daily life. In the following section, taking a product purchase guide as an example, we introduce the application of multilevel SEM.

A product purchase guide emerges due to the diversity of market products. Its purpose is to help the customer to select her products better, improve her degree of satisfaction in her purchased products, enhance the purchase efficiency, and achieve a win-win status between the customer and the seller. Analyzing the existing product purchase guides available on the market, we can categorize them in two main ways: promotion purchase guide and display purchase guide.

A promotion purchase guide is mainly a mall purchase guide, which actually promotes sales of products from the perspective of customers' experience, the point of which is to improve the customers' degree of satisfaction during the sales process and then create profit. However, such an approach lacks a systematic analysis between the products and the customers' needs.

A display purchase guide is mainly a network purchase guide, such as the TaoBao website. Such an approach carries out the transaction from the perspective of convenience: a convenient transaction and convenient management. Such an approach helps customers make their purchases by providing them with relatively complete product data and transaction information. Nevertheless, such a way lacks a comprehensive comparison of these data.

To summarize, the existing product purchase guide method can hardly yield a comprehensive and quantitative superiority evaluation of various products from the perspective of the customers' demands. To solve this problem, we use the multilevel SEM to perform a comprehensive and quantitative superiority evaluation for various products from the perspective of customers' demands supply customers with a more realistic and reliable selection basis.

Suppose we take the purchase that a middle-income customer makes of a China-made mobile phone as an example, and introduce the application of the multilevel SEM.

Determine the Measurable Indicators

The first-level measurable indicators of the mobile phones can be considered in the following aspects: In the economic aspect, the price can be regarded as an evaluation characteristic; in the technological aspect, various hardware parameters can be regarded as evaluation characteristics; in the social aspect, the online feedback information can be regarded as an evaluating characteristic; in the aspects of the combining economy and technology, the cost performance can be regarded as an evaluation characteristic. Thus, we obtain four first-level measurable indicators as follows:

$$MI = \left\{ MI_1, MI_2, MI_3, MI_4 \right\}, \text{Among them,}$$

$$MI_1 = (\text{Price}, \langle 700, 1500 \rangle \text{yuan});$$

$$MI_2 = (\text{Parameter satisfaction, acceptable range});$$

$$MI_3 = (\text{Cost performance}, >10);$$

$$MI_4 = (\text{Online feedback information, satistied})$$

$$MI \begin{cases} MI_1 & \\ MI_2 \begin{cases} MI_{21} \\ MI_{22} \\ MI_{23} \\ MI_{24} \end{cases} \\ MI_3 & \\ MI_4 \begin{cases} MI_{41} \\ MI_{42} \end{cases} \end{cases}, \alpha \begin{cases} \text{(Price)}\alpha_1 = 0.3 \\ \text{(Parameter)}\alpha_2 = 0.3 \begin{cases} \text{(Frequency of CPU)}\alpha_{21} = 0.4 \\ \text{(Running storage space)}\alpha_{22} = 0.3 \\ \text{(Battery capacity)}\alpha_{23} = 0.3 \\ \text{(Network requirment)}\alpha_{24} = \Lambda \end{cases} \\ \text{(Cost performance)}\alpha_3 = 0.2 \\ \text{(Online feedback information)}\alpha_4 = 0.2 \begin{cases} \text{(Sales valome of the whole month)}\alpha_{41} = 0.5 \\ \text{(Favorable rate)}\alpha_{42} = 0.5 \end{cases} \end{cases}$$

(a)   (b)

Figure 6.4.2   Measurable indicators and corresponding weight coefficients.

Because there are many mobile phone parameters and a great deal of online feedback information, MI_2 and MI_4 can be further divided:

1. By dividing the first-level indicator MI_2, we get the following corresponding second-level indicators:

$$MI_2 = \{MI_{21}, MI_{22}, MI_{23}, MI_{24}\}$$

$$MI_{21} = (\text{Frequency of CPU}, \langle 800, 2500 \rangle \text{ MHz}),$$

$$MI_{22} = (\text{Running storage space}, \geq 1 \text{ G}),$$

$$MI_{23} = (\text{Battery capacity}, \geq 1100 \text{ MAH}),$$

$$MI_{24} = (\text{Network requirment, support 3G network}).$$

2. By dividing the first-level indicator MI_4: $MI_4 = \{MI_{41}, MI_{42}\}$, we get the corresponding second-level indicators:

$$MI_{41} = (\text{Sales valome of the whole month}, \geq 10,000),$$

$$MI_{42} = (\text{Favorable rate}, \geq 90\%).$$

So, we obtain the measurable indicator system for evaluating the mobile phones made in China, as shown in Figure 6.4.2(a).

Determine the Weight Coefficients

Weigh coefficients denote the levels of relative importance of homologous indicators. The applicable objects of the product purchase guide are various hierarchies of customers, most of whom do not know how to determine the weight coefficients using scientific methods. To make multilevel SEM understandable, this case adopts a direct method by which the weight coefficients are determined according to customers' own demands, ensuring the generality and flexibility of the multilevel SEM in the application of product purchase guide. Figure 6.4.2(b) shows the weight coefficients of various measurable indicators. It can be seen that the weight coefficients satisfy the following laws:

1. MI_{24} denotes the feasibility of mobile 3G, which is a measurable indicator that must be satisfied. Therefore it should be used to perform the initial evaluation and has no weight.
2. The sum of the homologous weight coefficients is 1. As mentioned above, the values of the weighted coefficient satisfy:

$$\sum_{i=1}^{4} \alpha_i = 1, \quad \sum_{r_1=1}^{3} \alpha_{2r_1} = 1, \quad \sum_{r_2=1}^{1} \alpha_{4r_2} = 1,$$

Initial Evaluation

It can be known from the decision of weight coefficients that, for the two evaluation objects (China-made mobile phones S_1 and S_2), the condition that must be satisfied is the feasibility of mobile 3G. Certainly, both of them support the mobile 3G. We next continue the evaluation using the other measurable indicators.

Compute Dependence Degrees and Scaled Dependence Degrees

A dependent function is introduced to help customers select their products according to their own demands. Therefore, for an evaluating characteristic, the acceptable interval, satisfactory interval, and the combination of the acceptable interval and transition interval should be determined according to the customers' demands or other data. For the two types of mobile phones, dependence degrees of various characteristics can be computed as follows.

First, compute the dependence degree and scaled dependence degree about price.

The measurable indicator of price is $MI_1 = (\text{Price}, \langle 700, 1500 \rangle \text{ yuan})$.

It is well known that cheap mobile phones may have quality drawbacks, while expensive mobile phones can hardly be accepted by the majority of the public. We assume that the acceptable price range of the mobile phones is from 700 yuan to 1,500 yuan, that the relative satisfactory price range is from 900 yuan to 1,300 yuan, and that the ideal price is 1,200 yuan. When the price is between 500 yuan and 700 yuan, it can also be accepted if a promotion activity is performed or the demand of quality is not high. When the price is between 1,500 yuan and 2,000 yuan, it can also be accepted if the mobile phones have excellent quality and can be afforded. When the price is higher than 2,000 yuan or lower than 500 yuan, it is regarded as unacceptable. Thus, we can obtain the three intervals and the optimum of the dependence function as follows:

The optimal point: $x_{01} = 1200$;

Positive field:　$X_{01} = \langle a_{01}, b_{01} \rangle = \langle 900, 1300 \rangle$ yuan;

Standard positive field:　$X_1 = \langle a_1, b_1 \rangle = \langle 700, 1500 \rangle$ yuan;

Augmented field:　$\hat{X}_1 = \langle c_1, d_1 \rangle = \langle 500, 2000 \rangle$ yuan.

The online prices of S_1 and S_2 are 1,658 yuan and 998 yuan, respectively. The three intervals have no common endpoints, so we can obtain the dependence degrees of these two types of mobile phones using the elementary dependence function:

$$k_1(x_{11}) = \frac{\rho(x_{11}, x_{01}, X_1)}{D(x_{11}, x_{01}, X_1, \hat{X}_1)} = \frac{x_{11} - b_1}{(x_{11} - d_1) - (x_{11} - b_1)} = -0.316,$$

$$k_1(x_{12}) = \frac{\rho(x_{12}, x_{01}, X_1)}{D(x_{12}, x_{01}, X_{10}, X_1)} = \frac{a_1 - x_{12}}{(a_1 - x_{12}) - (a_{01} - x_{12})} = 1.49.$$

By scaling the dependence degrees, we can obtain the scaled dependence degrees as follows:

$$K_1(x_{11}) = \frac{k_1(x_{11})}{|k_1(x_{12})|} = \frac{-0.316}{|1.49|} = -0.212,$$

$$K_1(x_{12}) = \frac{k_1(x_{12})}{|k_1(x_{12})|} = \frac{1.49}{|1.49|} = 1.$$

Second, compute the dependence and scaled dependence degrees about mobile phones' parameters.

The measurable indicators of the mobile phones' parameters are $MI_2 = \{MI_{21}, MI_{22}, MI_{23}, MI_{24}\}$. Compute the dependence degrees according to the second-level indicators, respectively.

a. Central processing unit (CPU) indicator:　$MI_{21} = $ (Frequency of CPU, $\langle 800, 2500 \rangle$ MHz).

If the CPU's frequency is too high, it generates a lot of heat, consumes a lot of electricity, and has to meet other requirements of performance in other aspects. Thus, the CPU's frequency should be controlled in a suitable range, which is the same as the price analysis. The elementary dependence function is adopted and the three relevant intervals are determined as follows:

The optimal point: $x_{021} = 2,000$ MHz,

Standard positive field: $X_{021} = \langle a_{021}, b_{021} \rangle = \langle 1000, 2500 \rangle$ MHz;

Positive field: $X_{21} = \langle a_{21}, b_{21} \rangle = \langle 800, 3000 \rangle$ MHz;

Augmented field: $\hat{X}_{21} = \langle c_{21}, d_{21} \rangle = \langle 400, 3500 \rangle$ MHz.

Referring to the mobile phones' parameters, the CPU's frequencies of S_1 and S_2 are $x_{211} = 1,700$ MHz and $x_{212} = 1,300$ MHz, respectively.

Use the elementary dependence function to compute the dependence degrees about the CPU's frequency:

$$k_{21}(x_{211}) = \frac{\rho(x_{211}, x_{021}, X_{21})}{D(x_{211}, x_{021}, X_{021}, X_{21})} = 4.5,$$

$$k_{21}(x_{212}) = \frac{\rho(x_{212}, x_{021}, X_{21})}{D(x_{212}, x_{021}, X_{021}, X_{21})} = 2.5.$$

By scaling the dependence degrees, we obtain the scaled dependence degrees as follows:

$$K_{21}(x_{211}) = \frac{k_{21}(x_{211})}{|k_{21}(x_{211})|} = \frac{4.5}{|4.5|} = 1,$$

$$K_{21}(x_{212}) = \frac{k_{21}(x_{212})}{|k_{21}(x_{211})|} = \frac{2.5}{|4.5|} = 0.556.$$

b. Running inner storage indicator: $MI_{22} = $ (Running storage space, ≥ 1 G)

Running inner storage is a scatter of discrete points. So the discrete dependence function should be applied. The larger the running inner storage is, the better, and at least 1 gigabyte is acceptable. Namely, the dependence degree is 0 when the running inner storage is 1 gigabyte. According to the available data, the two types of mobile phones have equivalent running inner storages, both of which are 2 gigabytes. The discrete dependence function is:

$$k_{22}(x_{22}) = \begin{cases} 1, & x_{22} > 1 \text{ G} \\ 0, & x_{22} = 1 \text{ G} \\ -1, & x_{22} < 1 \text{ G}. \end{cases}$$

The scaled dependence degrees are $K_{22}(x_{221}) = K_{22}(x_{222}) = k_{22}(x_{221}) = k_{22}(x_{222}) = 1$.

c. Battery capacity indicator: MI_{23} = (battery capacity, ≥ 1100 MAH).

The battery capacities of S_1, S_2 are $x_{231} = 1{,}700$ MHz, $x_{232} = 1{,}300$ MHz, respectively. The dependence degrees of S_1, S_2 can be obtained by using the simple dependence function as follows:

$$k_{23}(x_{231}) = x_{231} - 1100 = (3050 - 1100) = 1950$$

$$k_{23}(x_{232}) = x_{232} - 1100 = (2300 - 1100) = 1200.$$

By scaling the dependence degrees, we obtain the following scaled dependence degrees:

$$K_{23}(x_{231}) = \frac{k_{23}(x_{231})}{|k_{23}(x_{231})|} = \frac{1950}{|1950|} = 1,$$

$$K_{23}(x_{232}) = \frac{k_{23}(x_{232})}{|k_{23}(x_{231})|} = \frac{1200}{|1950|} = 0.615.$$

Third, compute the dependence degrees and scaled dependence degrees of the cost performance indicator.

The cost performance indicator is MI_3 = (cost performance, >10). The cost performance can be denoted by the ratio of the performance score (obtained by the AnTuTu software) and the price. The larger the cost performance is, the better. The simple dependence function is adopted for use here:

S_1, S_2's performance scores and prices are respectively given as follows:

$$m_{31} = 35{,}100 \text{ score}, \quad m_{32} = 17{,}100 \text{ score},$$

$$m_{31}' = 1{,}658 \text{ yuan}, \quad m_{32}' = 998 \text{ yuan}.$$

Compute the cost performances of S_1 and S_2 respectively, as follows:

$$x_{31} = \frac{m_{31}}{m_{31}'} = \frac{35100}{1658} = 21.170,$$

$$x_{32} = \frac{m_{32}}{m_{32}'} = \frac{17100}{998} = 17.134.$$

So, we obtain their dependence degrees:

$$k_3(x_{31}) = x_{31} - 10 = 21.170 - 10 = 11.170$$

$$k_3(x_{32}) = x_{32} - 10 = 17.134 - 10 = 7.134.$$

By scaling the dependence degrees, we obtain the following scaled dependence degrees:

$$K_3(x_{31}) = \frac{k_3(x_{31})}{|k_3(x_{31})|} = \frac{11.170}{|11.170|} = 1,$$

$$K_3(x_{32}) = \frac{k_3(x_{32})}{|k_3(x_{31})|} = \frac{7.134}{|11.170|} = 0.639.$$

Finally, compute the dependence and scaled dependence degrees of the online feedback indicator.

The second-level indicators of the online feedback are $MI_4 = \{MI_{41}, MI_{42}\}$, where $MI_{41} = $ (Sales volume of the whole month, $\geq 10,000$) and $MI_{42} = $ (Favorable rate, $\geq 90\%$).

Monthly transaction amounts of S_1, S_2 are $x_{411} = 44,000$, $x_{412} = 3,525$, respectively. The larger the monthly transaction amount, the better. Use the simple dependence function to compute the dependence degrees of S_1, S_2:

$$k_{41}(x_{411}) = x_{411} - 10000 = 44000 - 10000 = 34000,$$

$$k_{41}(x_{412}) = x_{412} - 10000 = 3525 - 10000 = -6475.$$

By scaling the dependence degrees, we obtain the following scaled dependence degrees:

$$K_{41}(x_{411}) = \frac{k_{41}(x_{411})}{|k_{41}(x_{411})|} = \frac{34000}{|34000|} = 1,$$

$$K_{41}(x_{412}) = \frac{k_{41}(x_{412})}{|k_{41}(x_{411})|} = \frac{-6475}{|34000|} = -0.190.$$

The favorable ratings for S_1 and S_2 are 94% and 85%, respectively. The dependence degrees of S_1 and S_2 about the favorable rate are obtained by using the simple dependence functions $k_{42}(x_{42}) = x_{42} - 90\% = 0.8$, $k_{42}(x_{422}) = x_{422} - 90\% = -0.05$.

Therefore, we get the following scaled dependence degrees:

$$K_{42}(x_{421}) = \frac{0.04}{|-0.05|} = 0.8, \quad K_{42}(x_{422}) = -1.$$

Compute the Superiority

As mentioned above, we have three ways to compute the superiority. According to the specific situations given, a suitable computational method is selected. In this case, we adopt the weighted sum of comprehensive dependence degrees for the computation of the superiority. The superiority of object S is denoted by $C(S)$.

1. Because the measurable indicator MI_1 has no second-level indicators, we have

$$K_1(S_1) = K_1(x_{11}) = -0.212, \quad K_1(S_2) = K_1(x_{12}) = 1.$$

2. Because the first-level indicators MI_2, MI_4 can be decomposed into the following second-level indicators

$$MI_2 = \{MI_{21}, MI_{22}, MI_{23}, MI_{24}\}$$

where MI_{24} must be satisfied, we obtain the following comprehensive dependence degrees of the evaluating objects about measurable indicator MI_2:

$$K_2(S_1) = \sum_{r_1=1}^{3} K_{2r_1}(x_{2r_1 1}) \cdot \alpha_{2r_1} = 1$$

$$K_2(S_2) = \sum_{r_1=1}^{3} K_{2r_1}(x_{2r_1 2}) \cdot \alpha_{2r_1} = 0.7069.$$

3. Because the measurable indicator MI_3 has no second-level indicators, we have

$$K_3(S_1) = K_3(x_{31}) = 1, \quad K_3(S_2) = K_3(x_{32}) = 0.639.$$

4. The second-level indicators of measurable indicator MI_4 are $MI_4 = \{MI_{41}, MI_{42}\}$. So, the comprehensive dependence degrees of the evaluating objects about MI_{41} are:

$$K_4(S_1) = \sum_{r_4=1}^{2} K_{2r_4}(x_{2r_4 1}) \cdot \alpha_{2r_4} = 1 \times 0.5 + 0.8 \times 0.5 = 0.9$$

$$K_4(S_2) = \sum_{r_4=1}^{2} K_{2r_4}(x_{2r_4 2}) \cdot \alpha_{2r_4} = (-0.19) \times 0.5 + (-1) \times 0.5 = -0.595.$$

Therefore, we use the weighed sum to compute the S_1 and S_2 superiorities:

$$C(S_1) = \sum_{i=1}^{4} K_i(S_1) \cdot \alpha_i = 0.6164,$$

$$C(S_2) = \sum_{i=1}^{4} K_i(S_2) \cdot \alpha_i = 0.5209.$$

According to these results, mobile phone S_1 has a higher comprehensive superiority than mobile phone S_2. Under the adopted weight coefficients, it is more suitable to buy mobile phone S_1. Certainly, when selecting mobile phones, customers can alter the weight coefficients according to their own demands and preferences so that the obtained superiorities can reflect their individual characteristics.

6.4.3 Some Issues That Should Be Noted When Using SEM

So far, the SEM is one of the most widely applied methods, using dependence functions to determine the degrees of how much the evaluating object (basic-element) satisfies the requirement about a measurable indicator. For the practical requirement of the measurable indicator, we can select the simple dependence function, the elementary dependence function, or the discrete dependence function. We can also establish other types of dependence functions, although these must satisfy the establishment rules of dependence functions.

When it comes to multiple measurable indicators, we should compute the comprehensive superiority of the evaluating object according to the problem's practical requirements to judge whether the evaluating object is good or to determine its level of goodness. There are many methods that can be used to determine the weight coefficients, so it is important that an appropriate one is selected according to the specifics of the given problem.

The following issues should be noted when applying the SEM:

1. A dependence function should be appropriately selected; otherwise the results of the evaluation will be distorted.
2. When the value range of the evaluating object about a measurable indicator consists of nested intervals, these intervals are fixed in some cases and dynamic in other situations. If a dynamic evaluation is considered, it will be helpful to the researcher to use variable endpoints with hierarchically dynamic inner parts of the nested intervals.

Thinking and Exercises

1. By taking "level of English," "organizing ability," "language expression ability," "stature," and "appearance" as measurable indicators, simulate the recruiting requirements of an enterprise, and use the superiority evaluation method to carry out a comprehensive evaluation of the candidates.
2. Assume that you plan to rent an apartment in a city. Establish a first-level evaluating indicator system, and then use the one-level superiority evaluation method to comprehensively evaluate a particular apartment.
3. Assume that you plan to buy a residential rental building in a city. Establish a multilevel evaluating indicator system, and then use the multilevel superiority evaluation method to comprehensively evaluate the particular chosen building.

Chapter 7

Extension Creative Ideas Generating Methods for Solving Contradictory Problems

Content Summary

Extension set is a concept of sets proposed and gradually developed against the background of variable classification and a contradictory problem's quantitative judgement. It is a formalized quantitative tool that can be used to describe the contradictory problem and its converting process. It is also a set theoretic basis for judging contradictory degree and converting degree before and after a transformation is employed. Extension sets can be used to study variables' classification, cluster, recognition, control, and so on.

The extension creative idea generating method is developed based on the basic ideas of Extenics and uses a formalized and quantitative method to generate creative ideas for solving contradictory problems by imitating human thinking patterns. It establishes an extension model for the given contradictory problem based on the extension set theory, uses a dependence function to calculate the problem's compatible degree, performs extensible analysis, conjugative analysis, and extension transformation on the problem, and obtains superior extension creative ideas of solving the problem via superiority evaluation.

In this chapter, we introduce the basis knowledge of extension sets, then we introduce the extension creative idea generating method for solving contradictory problems.

7.1 Pre-knowledge

- The variable classification problem: To check if some workpieces are qualified by the classification method of Cantor set, the workpieces can be regarded as either qualified or unqualified according to their diameters. For the unqualified workpieces, if we adopt a reprocessing approach, those workpieces with diameters larger than the requirement can be reprocessed, those with diameters smaller than the requirement are waste; however, if we adopt an electroplating approach, those with diameters smaller than the requirement can be reprocessed, and the others are waste. Thus, workpieces that can be reprocessed are special transformation-based unqualified workpieces.
- How can one change unqualified products to qualified? Find a solution to the problem, namely a transformation.
  How can one measure the quantitative degree by which an unqualified product changes to a qualified one? Establish such a function of the characteristic "diameter" by which workpieces satisfy the requirement.
- Such problems belong to the category of problems of variable classification, which cannot be described by Cantor set (deterministic classification) or fuzzy sets (fuzzy classification).

7.1.1 Comparison of Three Types of Sets

The concept of sets is the basis for the human brain to classify and recognize things.

Cantor set: Regard an object as having property p or not having property p. This property is described by the feature function with the value range $\{0, 1\}$.

Fuzzy set: Regard an object as having property p by various membership degrees ranging over the interval $[0, 1]$.

Extension set: Regard an object as having property p, not having property p, or being in the critical status, which is described by a dependence function defined on the entire real number field. Extension set can also describe the dynamic conversion among the three situations above.

In Extenics, extension set is applied to the quantitative description of things' variation and dynamic classification. It is the basis of solving contradictory problems and a formalized description of qualitative and quantitative variation.

These three types of sets are illustrated by Figure 7.1.1.

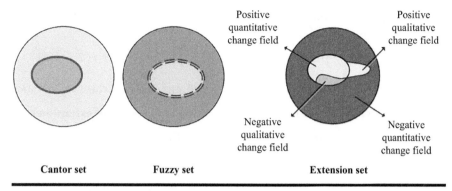

Cantor set　　　　Fuzzy set　　　　　　　Extension set

Figure 7.1.1　Comparison of three types of sets.

7.1.2 Definition of Extension Sets

To understandably introduce the concept of extension sets and function, we take the recruitment of an enterprise as an example. By proposing a problem, we then provide the definition of extension sets and use it to solve the problem.

PROBLEM 1

For all candidates, how can they be described in a formalized and quantitative way as those who are qualified (satisfying the recruiting requirements), those who are unqualified (not satisfying the recruiting requirements), and those who are in the critical status?

PROBLEM 2

Can unqualified candidates be reclassified as qualified? Can qualified candidates be changed to be unqualified? How can a formalized and quantitative description be given to describe the qualitative change?

PROBLEM 3

Are there any candidates who are qualified both before and after the transformation? Are there any candidates unqualified both before and after the transformation?

PROBLEM 4

Are there any candidates in the critical status after the transformation?

PROBLEM 5

What transformations can be implemented for the qualitative change or quantitative change?

To solve these problems, we give the following definition of extension sets.

Definition 7.1.1 Let U be the domain of all candidates recruited by the company of concern, let $u \in U$ be a candidate, let $y = k(u)$ be the degree of how much candidate u satisfies the recruiting requirements, and k a dependence criterion. Then the extension set of the domain U is

$$\tilde{E}(T) = \left\{(u, y, y') \,\middle|\, u \in U, y = k(u) \in \mathfrak{R}; \quad T_u u \in T_U U, y' = T_k k(T_u u) \in \mathfrak{R}\right\},$$

where $T = (T_U, T_k, T_u)$ is a transformation applied, and $\mathfrak{R}$ is the field of real numbers.

Now let's solve the above listed problems using the concept of extension sets.

1. When no transformation is applied, the set of all qualified candidates is denoted by

$$E_+ = \left\{(u, y) \,\middle|\, u \in U, y = k(u) > 0\right\},$$

which is called the positive field of $\tilde{E}$. The set of unqualified candidates is denoted by

$$E_- = \left\{(u, y) \,\middle|\, u \in U, y = k(u) < 0\right\},$$

which is called the negative field of $\tilde{E}$.

The set of candidates in the critical status (for example, someone's qualification of a certificate has been approved, but the certificate has not yet been awarded) is denoted by

$$E_0 = \left\{(u, y) \,\middle|\, u \in U, y = k(u) = 0\right\},$$

which is called the zero field of $\tilde{E}$.

Using these three sets, a formal quantitative description of Problem 1 has been solved.

2. Suppose that the recruitment has certain requirements of computer skills. When the domain U and the dependence criterion k are constant, let T_u denote the transformation of performing computer operating training to the candidates for a week. Some candidates have improved their computer skills after the training. The set of candidates who are unqualified before the training but are qualified after the training is denoted by

$$E_+(T_u) = \left\{(u, y, y') \,\middle|\, u \in U, y = k(u) \leq 0; T_u u \in U, y' = k(T_u u) > 0\right\}$$

which is called the positive extension field or the positive qualitative change field of $\tilde{E}(T)$. These candidates are changed to being qualified, but because the number of employment openings is limited, some originally qualified people now become unqualified due to their relatively low computer skills. Jointly, the set of these candidates is denoted by

$$E_-(T_u) = \left\{(u, y, y') \,\middle|\, u \in U, y = k(u) \geq 0; T_u u \in U, y' = k(T_u u) < 0\right\},$$

which represents all the candidates who were qualified previously but washed out afterwards. This set is called the negative extension field or the negative qualitative change field of $\tilde{E}(T)$.

Using these two sets just defined, the formal quantitative description as mentioned in Problem 2 is solved.

3. In (2), the set of candidates who are qualified before and after the transformation T_u is denoted by

$$E_+(T_u) = \left\{(u, y, y') \mid u \in U, y = k(u) > 0; \, T_u u \in U, y' = k(T_u u) > 0\right\},$$

which is called the positive stable field or the positive quantitative change field of $\tilde{E}(T)$. The set of candidates who are unqualified before and after the transformation T_u is denoted by

$$E_-(T_u) = \left\{(u, y, y') \mid u \in U, y = k(u) < 0; \, T_u u \in U, y' = k(T_u u) < 0\right\},$$

which is called the negative stable field or the negative quantitative change field of $\tilde{E}(T)$.

Using these two sets, the formal quantitative description as mentioned in Problem 3 is solved.

4. In (2), the set of candidates who are in the critical status after the transformation T_u is denoted by

$$E_0(T_u) = \left\{(u, y, y') \mid u \in U, T_u u \in U, y' = k(T_u u) = 0\right\},$$

which is called the extension field of $\tilde{E}(T)$.

Using this set, the formal quantitative description as mentioned in Problem 4 is solved.

The objects of the above transformations are elements in the domain, namely candidate u. According to the definition of extension sets, domain U and dependence criterion k are also variable, of which the transformations can lead to changes that make candidates either satisfy or fail to satisfy the requirements.

5. If the domain and candidates in the domain all stay unchanged and T_k denotes the transformation of the dependence criterion k, the corresponding extension set is

$$\tilde{E}(T_k) = \left\{(u, y, y') \mid u \in U, y = k(u) \in \Re, y' = T_k k(u) \in \Re\right\}.$$

Suppose that T_k alters some of the recruiting requirements, such as decreasing the requirement of educational experience, strengthening the requirement of linguistic expressing ability. Then the positive qualitative change field $E_+(T_k)$ denotes the candidates who were unqualified originally but changed to be qualified after the transformation of recruiting requirements; the negative qualitative change field $E_-(T_k)$ denotes the candidates who were qualified originally but change to be unqualified after the transformation of recruiting requirements; the positive quantitative change field $E_+(T_k)$ denotes the candidates who are qualified before and after the transformation of recruiting requirements; the negative quantitative change field $E_-(T_k)$ denotes the candidates who were unqualified both before and after the transformation of recruiting requirements.

6. If T_U denotes the transformation of the domain U, the corresponding extension set is

$$\tilde{E}(T_U) = \left\{(u, y, y') \mid u \in U, y = k(u) \in \Re; u \in T_U U, y' = k'(u) \in \Re\right\},$$

where

$$k'(u) = \begin{cases} k(u), & u \in U \cap T_U U \\ k_1(u), & u \in T_U U - U \end{cases}.$$

Namely, when candidates are in the intersection of the original domain U and the renewed domain $T_U U$, the recruiting requirements stay constant; when some candidates move out of the original domain U, we should redefine the recruiting requirements (new recruiting requirements can also be same as the original ones); namely, both $k_1(u) = k(u)$ and $k_1(u) \neq k(u)$ are acceptable.

Suppose that T_U expands the recruiting area, and all other recruiting requirements stay constant. For example, suppose that the original recruiting area was Beijing, namely the domain $U =$ {working-age candidates in Beijing}; now, we expand U to $T_U U =$ {working-age candidates in China}. The positive qualitative change field denotes all candidates out of Beijing satisfying all other recruiting requirements; the negative qualitative change field denotes candidates from Beijing who are originally qualified but change to be unqualified due to the addition of candidates from outside of Beijing; the positive quantitative change field denotes candidates who stay qualified before and after the transformation of recruiting requirements; the negative quantitative change field denotes candidates who stay unqualified before and after the transformation of recruiting requirements.

In particular, when $T_u = e$, $T_k = e$, and $T_U U \subset U$, $T_k k = k$, $T_u u = u$, $y' = k(u) = y$, and

$$\tilde{E}(T) = \tilde{E}(T_U) = \{(u, y) \mid u \in T_U U, y = k(u) \in \Re\}.$$

Therefore, the concept of extension sets can provide formalized and quantitative expressions of conversions of objects' properties, and by using extension sets people can study the objects' variable classification.

These statements imply that Problem 5 is solved: The qualitative or quantitative change can be implemented by a transformation of the elements in the domain, dependence criterion, or the domain.

Considering a transformation T of object u, the division of the domain performed by the extension set is illustrated by Figure 7.1.2.

According to above definitions, an extension set describes the mutual conversion of matters and affairs between "yes" or "no." It can be used to describe not only the quantitative process (quantitative field) but also the qualitative process

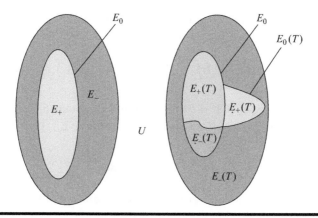

Figure 7.1.2 **Apply a transformation T on object u, the division of the domain performed by the extension set.**

(qualitative field). The zero field or the extension field is the boundary of the qualitative changes, and beyond them, things have a qualitative change.

The core concept of extension sets is the qualitative field. It consists of a positive qualitative change field and a negative qualitative change field. Different transformations correspond to different qualitative fields. The elements of the qualitative field have experienced qualitative changes through the applied transformation. The introduction of qualitative fields paves a reasonable theoretical path to converting given contradictory problems to non-contradictory problems.

7.1.3 Basic-element Extension Set

A basic-element extension set is a result of combining basic-element theory and extension set theory. In a basic-element extension set, the elements of the domain can be matter-element, affair-element, or relation-element. Each basic-element has an inherent structure, and transformations of its three essential factors can alter its position in the extension set. Therefore, basic-element extension sets can be a quantitative tool to describe the variability of matters and affairs.

For some problems, multiple evaluation characteristics are required to evaluate the elements (basic-elements) of the domain. We need to establish a multiple-evaluation characteristic basic-element extension set. It is the theoretical basis of multiple-indicator comprehensive evaluation (superiority evaluation) and multiple-characteristic incompatible problem solving.

Definition 7.1.2 For a basic-element set $S = \{B\}$, suppose that $c_0 = (c_{01}, c_{02}, ..., c_{0m})$ is the m evaluation characteristics of B, and the value of B about c_0 is

$$c_0(B) = \left(c_{01}(B), c_{02}(B), ..., c_{0m}(B)\right) \triangleq \left(x_1, x_2, ..., x_m\right).$$

$V(c_{0i})$ is the value field of x_i, and X_{0i} is the positive field, $X_{0i} \subset V(c_{0i})$. We establish dependence functions $k_i(x_i)$, $i = 1, 2, ..., m$. We denote:

$$k\left(c_0(B)\right) = \left(k_1\left(c_{01}(B)\right), k_2\left(c_{02}(B)\right), ..., k_m\left(c_{0m}(B)\right)\right) = \left(k_1(x_1), k_2(x_2), ..., k_m(x_m)\right)$$

as the evaluation vector of B. If the corresponding weight coefficients of the problem of concern are $\alpha_1, \alpha_2, ..., \alpha_m$, respectively, satisfying $\sum_{i=1}^{m} \alpha_i = 1$, we call

$$K(B) = \sum_{i=1}^{m} \alpha_i k_i\left(c_{0i}(B)\right) = \sum_{i=1}^{m} \alpha_i k_i(x_i)$$

the comprehensive dependence degree of B about c_0, and we call

$$\tilde{E}(B)(T) = \left\{(B, Y, Y') \mid B \in S, Y = K(B) \in \Re; T_B B \in T_S S, Y' = T_K K(T_B B) \in \Re\right\}$$

the multiple-evaluating characteristic basic-element extension set on S.

For comprehensive dependence degrees, we have the following special situations:

1. If a basic-element is regarded to be qualified only if it is qualified about each evaluation characteristic, the comprehensive dependence degree is defined as

$$K(B) = \bigwedge_{i=1}^{m} k_i\left(c_{0i}(B)\right) = \bigwedge_{i=1}^{m} k_i(x_i)$$

where $\bigwedge_{i=1}^{m}$ denotes the minimum of the m values.

2. If a basic-element is regarded to be qualified only if it qualifies about at least one evaluation characteristic, the comprehensive dependence degree is defined as:

$$K(B) = \bigvee_{i=1}^{m} k_i\left(c_{0i}(B)\right) = \bigvee_{i=1}^{m} k_i(x_i)$$

where $\bigvee_{i=1}^{m}$ denotes the maximum of the m values.

3. If the value range of a characteristic c_{0i_0} must be satisfied, we first use the characteristic to filter through all the basic-elements, and for the satisfied basic-elements we establish a comprehensive dependence function by using the rest of the evaluation characteristics.

In a multiple-evaluating characteristic basic-element extension set, each basic-element B_j ($j = 1, 2, ..., n$) is in domain S if $K(B_j) > 0$, B_j is regarded as qualified; if $K(B_j) < 0$, B_j is regarded as unqualified; if $K(B_j) = 0$, B_j is regarded as being in the critical status, which needs to be further analyzed according to the definite conditions, because some practical problems regard the critical status as being qualified, while other problems regard it as being unqualified.

If there is a basic-element B_0 that satisfies

$$K(B_0) = \max_{1 \le j \le n}\left\{K(B_j), B_j \in S\right\}$$

then B_0 has the largest comprehensive dependence degree, namely the largest superiority (the degree of how much it satisfies the requirement), which can supply a quantitative evidence for processing contradictory problems.

Case Analysis

A university categorizes its students for convenient administration. The original categorizing method is according to academics, major, educational experience, age, and so on. Modern society's development makes some academics and majors more and more intimate, so the variability should be considered in the category of students. In the following section, we use basic-element extension sets to give several extension categorizing methods for student categorization.

Suppose that domain U denotes all students in this university. Each student in U can be expressed through using matter-element M:

$$
M = \begin{bmatrix}
O, & \text{academy } c_1, & v_1 \\
& \text{major } c_2, & v_2 \\
& \text{graduate category } c_3, & v_3 \\
& \text{speciality } c_4, & v_4 \\
& \text{age } c_5, & v_5 \\
& \text{professional ability } c_6, & v_6
\end{bmatrix}
$$

We have $M \in U$.

The university organizes its students to carry out a new research project in their spare time. The project requires the students' professional ability. Let $y = k(M)$ describe the quantitative degree of how much a student satisfies the requirement. Thus, we can establish the following matter-element extension set:

$$
\tilde{E}(T) = \left\{ (M, y, y') \mid M \in U, y = k(M) \in \Re; T_M M \in T_U U, y' = T_k k(T_M M) \in \Re \right\}.
$$

1. When no transformation is performed, namely $T = (T_k, T_M, T_U) = (e, e, e)$, all students can be categorized into three categories:

$E_- = \{M \mid M \in U, k(M) < 0\}$ denotes the set of students whose professional ability doesn't satisfy the requirement;

$E_+ = \{M \mid M \in U, k(M) > 0\}$ denotes the set of students whose professional ability satisfies the requirement; and

$E_0 = \{M \mid M \in U, k(M) = 0\}$ denotes the set of students in the critical status.

By the above categorization, the number of qualified students is insufficient to complete the project. Therefore, the unqualified students have to be trained. Thus, we need to perform transformation-based categorization to obtain a new research team.

2. If domain U and dependence criterion k stay constant, namely $T_U U = U$, and $T_k k = k$, perform technical training on unqualified students, namely performing transformation:

$$
T_M M = \begin{bmatrix} O, & \text{academy } c_1, & v_1 \\ & \text{major } c_2, & v_2 \\ & \text{graduate category } c_3, & v_3 \\ & \text{speciality } c_4, & v_4 \\ & \text{age } c_5, & v_5 \\ & \text{professional ability } c_6, & v_6' \end{bmatrix} = M'.
$$

After training, a corresponding exam is carried out to determine the degree of how much the students now satisfy the requirement. According to the value of $y' = k(T_M M)$, the students can be categorized into four categories:

$E_+ = \{M \mid M \in U, k(M) > 0\}$ denotes the set of qualified students before the training;

$E_+(T) = \{M \mid M \in U, y = k(M) \le 0; T_M M \in U, y' = k(T_M M) > 0\}$, denotes the set of unqualified students or the students in the critical status before the training, who, however, change to be qualified after the training;

$E_-(T) = \{M \mid M \in U, y = k(M) < 0; T_M M \in U, y' = k(T_M M) < 0\}$, denotes the set of students who stay unqualified both before and after training; and

$E_0(T) = \{M \mid M \in U, T_M M \in U, y' = k(T_M M) = 0\}$, denotes the set of students in the critical status after the training, no matter whether or not they originally met the requirements.

Under this transformation T_M, the set of qualified students is changed to $E_+ \cup E_+(T)$. If the number of these students is sufficient to complete the project, the categorization is over. If the number of qualified students is more than needed, sort them according to their values of the dependence function and select the superior ones.

3. If the number of qualified students is still insufficient to complete the project, the transformation of dependence criterion can be considered, namely performing a transformation $T_k k = k'$ to decrease the requirement of professional ability and increase the requirement of learning ability so that some initially unqualified students can become competent to do the work through learning after entering the team.

4. If the personnel requirement still cannot be satisfied by this categorization, a transformation of the domain can be considered, namely performing transformation $T_U U = U'$, such as recruiting qualified students from other universities. That can also make the categorization change and generate a new team.

7.2 Defining Methods and Extension Models for Problems

QUESTIONS AND THINKING:

Question 1. In the Three Kingdoms' time period, someone gave Caocao an elephant. Caocao wondered at the elephant's weight, which was far beyond the maximum measurement (200 kg) of the scale used at that era. How could one weigh the elephant? Is Caochong's solution for weighing elephant the best? How can one establish a model for the problem? How can she obtain creative ideas of solving the problem?

Question 2. The traffic rule of Hong Kong is to drive on the left side of the road, while the traffic rule of Shenzhen is to drive on the right side of the road. How can one connect the two places without altering the traffic rules in Hong Kong and Shenzhen? How can you establish a model for the problem? How can you obtain creative ideas for solving the problem?

- In these problems, one or multiple goals cannot be implemented. Such problems are called contradictory problems. Are there any laws that govern the solution of contradictory problems? Are there any methods to follow?
- These problems cannot be expressed by conventional mathematical models. Can we establish a formalized quantitative model to describe contradictory problems?

7.2.1 Defining Method of a Problem

When encountering a contradictory problem, we should first define it appropriately, which is the basis of solving it. To define a problem, its goal and condition should be defined. To define the goal, it should be made definite, namely making the goal marked by some approach. Clarifying and quantifying the goal can increase the probability of resolving the problem.

7.2.1.1 Defining the Goal

A goal is the basis of actions. A goal generates belief. Clear goals generate rigid belief, while vague goals can hardly be worked on. A clear and accurate goal is the essential premise of successfully solving a problem or generating some effects. It is also the basic evidence of evaluating decision-making schemes and developing results. Thus, a definite goal makes evaluation of the consequence easier.

Setting the goal is crucial to finding clues and creative ideas that will lead to the eventual solution to the problem. The goal is also an indicator for selecting highly feasible and efficient schemes. The goal has to be formalized and digitalized. Only in this way can the goal be genuinely definite, and only in this way can a model for the problem be established appropriately. If the goal is unclear, it is hard to determine the essential factors, tools, and personnel allocation to complete the intended task.

In practice, there are often multiple goals. At such situations, the implicating system method introduced in Chapter 3 should be used. Perform an implicating analysis on the goals to determine the hierarchy among them. For the goals in a same hierarchy, the order should also be determined. If the most superior goal is g, the implicating system is:

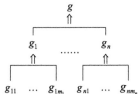

Implementing goals should proceed step by step from a low level to a high level, while setting goals should move from a high level to a low level. All the goals form a goal-implicating system. Implementation of inferior goals implicates the implementation of superior goals. Goals in a same hierarchy may have some correlations.

If there is only one superior goal, it is a single-goal problem; if there are multiple superior goals, it is a multiple-goal problem. The implicating system of multiple-goal problem is illustrated by the following figure.

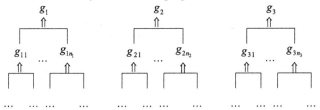

Goals include single-goal and multiple-goal, staged goal and long-term goal, local goal and global goal. Therefore, at the beginning of solving the problem of concern, we should clearly know the goals and the relationship among them.

7.2.1.2 Defining the Condition

After defining the goals, the conditions should be analyzed and defined. Conditions involve resource conditions and environmental conditions. Resource conditions involve internal resources and external resources. Environmental conditions involve internal environment and external environment. Conditions are mostly objective, but also creatable and transformable. Among most conditions, some are beneficial to implementing the goals, while others are harmful; some are compatible with the goals, while others are contradictory to the goals; some are non-limiting conditions, and others are limiting conditions. All conditions must be clearly defined.

The procedure of defining conditions consists of

1. Collect goal-related data.
2. Analyze conditions $l_i (i = 1, 2, \cdots, n)$ required for implementing the goals.
3. Trim l_i-related data, and determine the practical conditions l_i' corresponding to l_i.
4. Analyze the difference between l_i' and l_i to determine whether l_i''s are limiting conditions or non-limiting conditions.
5. Express l_i' using basic-elements; define the matters, affairs, actions, characteristics, and values that the conditions involve.
6. Select primary goal-related beneficial conditions and limiting conditions.

An important procedure of defining conditions is to express the conditions using basic-elements. In this way the conditions can be digitalized so that further analysis and goal implementation can be accomplished by looking at the conditions or by extending the conditions.

Sometimes people are limited in achieving their goals due to some objective existence or limitations of human-defined conditions. To determine limiting conditions, the limiting property should be noted: elastic limit or rigid limit, implicated limit or dominant limit, uncertain limit or certain limit. Only in this way can the conditions be analyzed correctly.

7.2.1.3 Defining the Problem

Appropriately defining the problem (simplifying, clarifying, and formalizing the problem) and finding the key problem is equivalent to solving half of the problem. The problem is constituted by goals and conditions. After defining the goals and conditions, the problem can be formalized by using basic-elements or compounded-elements.

We usually concern ourselves with the contradiction between the goals and conditions, namely, the incompatible problem. In such a time, the goals do not seem to lead to a consistent conclusion. If two different goals are considered under the same conditions or if there are contradictions between the goals, it is an antithetical problem.

A single-goal problem, formed by G and L, is denoted by $P = G * L$. So, the problem is defined. The next step of the procedure is to judge if the problem is a contradictory problem.

For a problem of multiple goals and multiple conditions, we need to express the relationships between the goals and conditions, the relationships among the goals, and the relationships among the conditions using basic-elements or compounded-elements, followed by the next step, which is to establish a model for the problem.

7.2.2 The Problem's Extension Model

Any problem is constituted with goals and conditions, which can be expressed using either basic-elements or compounded-elements. Therefore, any problem can be expressed by a formalized model using basic-elements or compounded-elements.

If the goals cannot be implemented under the given conditions, it is called a contradictory problem; otherwise, it is called a non-contradictory problem. We are concerned about how to solve a contradictory problem. To do so, we must clarify goals and conditions and define the problem clearly. The next step is to establish a model for the problem so that the contradictory problem can be analyzed and solved more clearly, qualitatively and quantitatively.

In the following section, we introduce the methods of establishing an extension model for problems with different types of goals.

7.2.2.1 Single-goal Problem's Extension Model

Suppose that problem P's goal and condition are respectively G and L (where L can be either a single condition or a set of multiple conditions), which can both be expressed by formalized models using either basic-elements or compounded-elements. Model $P = G * L$ is called an extension model of the single-goal problem.

If goal G can't be achieved under the condition L, P is an incompatible problem, denoted as $G \uparrow L$. The methods of judging and solving the incompatible problem will be introduced in Section 7.3.

Example 7.2.1

A high-tech enterprise E has 1,000 staff members and a good reputation. The enterprise plans to hold a workshop W with an approximate cost of <4,500,000, 5,000,000> yuan. The enterprise's current available funds, however, total no more than 1,000,000 yuan. An extension model of the problem is

$$P = G * L,$$

$$G = \begin{bmatrix} \text{Establish,} & \text{dominating object,} & \text{workshop } W \\ & \text{acting object,} & \text{enterprise } E \\ & \text{cost,} & < 4,500,000, 5,000,000 > \text{ yuan} \end{bmatrix},$$

$$L = \begin{bmatrix} \text{Enterprise } E, & \text{amount of available capital,} & 1,000,000 \text{ yuan} \\ & \text{amount of staff,} & 1,000 \\ & \text{project type,} & \text{high technology} \\ & \text{reputation,} & 5 \end{bmatrix}.$$

7.2.2.2 Dual-goal Problem's Extension Model (the two goals should be achieved simultaneously)

Suppose that problem P's goals are G_1 and G_2, which need to be achieved simultaneously. P's condition is L (which can be a single condition or a set of multiple conditions). The goals G_1 and G_2 and the condition L can all be expressed into formalized models using either basic-elements or compounded-elements. The model

$$P = (G_1 \wedge G_2) * L$$

is called the dual-goal problem's extension model (the two goals should be achieved simultaneously).

If goals G_1 and G_2 can't be achieved simultaneously under the condition L, then P is an antithetical problem, denoted as $(G_1 \wedge G_2) \uparrow L$. The methods of judging and solving an antithetical problem will be detailed in Section 7.4.

The dual-goal multiple-condition problem's extension model (the two goals should be achieved simultaneously) is denoted as

$$P = (G_1 \wedge G_2) * (L_i, \wedge, \vee, \neg), \quad i = 1, 2, \cdots, n.$$

Example 7.2.2

Put a wolf O_1 and a chicken O_2 in the same cage and ensure that the chicken is not eaten by the wolf. The problem's extension model is

$$P = (G_1 \wedge G_2) * L$$

where

$$G_1 = \begin{bmatrix} O_1, & \text{habit}, & \text{eating meat} \\ & \text{location}, & \text{cage } O \end{bmatrix} = \begin{bmatrix} O_1, & c_1, & v_{11} \\ & c_2, & v_{12} \end{bmatrix}$$

$$G_2 = \begin{bmatrix} O_2, & \text{habit}, & \text{temperate} \\ & \text{location}, & \text{cage } O \end{bmatrix} = \begin{bmatrix} O_2, & c_1, & v_{21} \\ & c_2, & v_{22} \end{bmatrix}.$$

$$L = \left(\text{Cage } O, \quad \text{volume}, \quad a\,\text{m}^3 \right) = (O, \quad c, \quad v)$$

7.2.2.3 Dual-goal Problem's Extension Model with at Least One Goal Achieved

Suppose that problem P's goals are G_1 and G_2, at least one of which needs to be achieved. P's condition is L (which can be a single condition or a set of multiple conditions). G_1, G_2, and L can all be expressed by formalized models using either basic-elements or compounded-elements. The model $P = (G_1 \vee G_2) * L$ is called a dual-goal problem's extension model with at least one goal achieved.

Such a dual-goal multiple-condition problem extension model (with at least one goal achieved) is denoted as

$$P = (G_1 \vee G_2) * (L_i, \wedge, \vee, \neg), \quad i = 1, 2, \cdots, n.$$

Such a problem can usually be decomposed into incompatible problems to solve.

7.2.2.4 Multiple-goal Problem's Extension Model (with the goals achieved simultaneously)

If problem P has multiple goals that need to be achieved simultaneously, the extension model can be expressed as:

$$P = (G_1 \wedge G_2 \wedge \cdots \wedge G_m) * (L_i, \wedge, \vee, \neg), \quad i = 1, 2, \cdots, n.$$

7.2.2.5 General Multiple-goal Problem's Extension Model

Generally, if problem P has multiple goals, its extension model can be expressed as

$$P = (G_i, \wedge, \vee, \neg) * (L_j, \wedge, \vee, \neg)$$

where $\wedge, \vee, \neg$ denote the operators "and," "or," and "not," respectively, among the goals or conditions.

7.3 Extension Creative Idea Generating Method for Solving Incompatible Problems

QUESTIONS AND THINKING:

- Have you heard about the story of Caochong weighing elephant? How can Caochong weigh a big elephant using a small scale? Do you have a better method than that used by Caochong?
- Suppose you want to buy a 100-square meter house, which has a price of 1,000,000 yuan, but you have only 200,000 yuan. How do you plan to purchase the house?
- The incompatible problem is the most common type of contradictory problems. When you regard a problem as an incompatible problem, what is the deciding evidence? How do you actually make this determination?

Based on the extension models introduced in the last section, we first must use the concepts of extension sets and dependence functions to present how to establish the incompatible problem's and its core problem's extension models and how to produce the necessary quantitative judgment. Then we introduce an extension creative idea generating method of solving the incompatible problem.

A creative idea stands for extension transformations or extension transformation operations that can make the compatible degree of the incompatible problem from less than zero or zero to larger than zero, namely the problem's solution transformation. The process of generating extension creative ideas is called extension creative idea generation (ECIG).

7.3.1 Judging Method of the Incompatible Problem

Consider the problem $P = G * L$, where G and L are basic-elements or compounded-elements. If the goal G can't be achieved under the condition L, then problem P is called an incompatible problem, denoted as $G \uparrow L$.

We have three paths to solving the incompatible problem: (1) Keep the goal constant and solve the problem by altering the condition; (2) keep the condition constant and solve the problem by altering the goal; and (3) solve the problem by altering both the goal and the condition.

When it is difficult or seemingly impossible to solve a problem, first define its goal and condition and establish the problem's extension model; then judge which of them—goal or condition—to alter; determine the evaluation characteristics and establish the core problem's extension model; and finally establish the core problem's compatible degree function to judge the problem's compatible degree.

1. When the goal is invariable, try to alter the condition to solve the incompatible problem. At this time, a compatible degree function must be established by taking the measurement value range in which the goal can be achieved as the positive field. Determine the incompatible degree of the problem.
2. When the condition is invariable, try to alter the goal to solve the incompatible problem. At this time, a compatible degree function must be established by taking the measurement value range. The condition can supply as the positive field. Determine the incompatible degree of the problem.
3. When the goal and condition are both invariable, perform an implicating analysis on the goal to find its inferior goals. If the inferior goals are variable, judge the problem's incompatible degree according to (2); if the inferior goals cannot be found, perform an implicating analysis on the condition to find its inferior conditions and judge the problem's incompatible degree according to (1).
4. When the goal and condition are both variable, an incompatible degree function must be established, usually by taking the measurement value range in which the goal can be achieved as the positive field. Determine the incompatible degree of the problem.

The general judging method of the incompatible problem is stated as follows:

Given problem $P = G * L$, where G and L are either basic-elements, compounded-elements, or a basic-elements operation. Suppose that c_0 is the evaluation characteristic; c_{0s} is the characteristic required by object Z the problem involves; the positive is X_0; c_{0t} is the characteristic another problem-involving object Z_0 can supply, and the value is x_0. We denote this as

$$g_0 = (Z, \quad c_{0s}, \quad X_0), \quad l_0 = (Z_0, \quad c_{0t}, \quad x_0),$$

and $P_0 = g_0 * l_0$ is called problem P's core problem.

Taking X_0 as the positive field, establish a compatible degree function of object Z about c_0, namely, problem P's compatible degree function $k(x)$ (see the method of establishing dependence function in Section 6.1). Denote $K_0(P) = k(x_0)$, called problem P's compatible degree. If $K_0(P) < 0$, problem P is an incompatible problem; if $K_0(P) > 0$, problem P is a compatible problem; if $K_0(P) = 0$, problem P is a critical problem.

Case Analysis:

Example 7.3.1

Person E wants to buy a 130-square meter apartment W in a building with twelve floors. Because different floors have different prices, the price range is [2,500,000, 3,000,000] yuan. Presently his family's monthly income is 40,000 yuan, his available savings is 1,200,000 yuan, his ideal floor is the sixth floor, and the price of apartments on this floor is 2,800,000 yuan. Suppose that the goal of buying an apartment stays constant, and we can only solve the problem by altering the condition. Establish the problem's extension model and judge its compatible degree.

Solution: The problem's extension model is

$$P = G * L$$

$$= \begin{bmatrix} \text{Buy,} & \text{dominating object,} & M \\ & \text{acting object,} & E \end{bmatrix} * \begin{bmatrix} E, & \text{amount of available capital,} & 1,200,000 \text{ yuan} \\ & \text{monthly household income,} & 40,000 \text{ yuan} \\ & \text{profession,} & \text{white collar workers} \\ & \text{reputation,} & 5 \end{bmatrix}$$

where

$$M = \begin{bmatrix} W, & \text{number of floors,} & 6 \text{ floor} \\ & \text{price,} & 2,800,000 \text{ yuan} \end{bmatrix}.$$

Suppose that the evaluation characteristic is "savings." We can denote this as the supplied funds $= c_0$, and the core problem's extension model is

$$P_0 = g_0 * l_0$$

$$= (W, \quad c_{0s}, \quad [2,500,000, \quad 3,000,000] \text{ yuan}) * (E, \quad c_{0r}, \quad 1,200,000 \text{ yuan}).$$

Taking $x_0 = 2,800,000$ yuan as the optimum point of required funds, the positive field is $X = [2,500,000, \quad 3,000,000]$; establish a compatible degree function according to the simple dependence function introduced in Chapter 6:

$$k(x) = \begin{cases} \dfrac{x - 2500000}{2800000 - 2500000}, & 0 \le x \le 2800000 \\ \dfrac{3000000 - x}{3000000 - 2800000}, & x \ge 2800000 \end{cases}$$

$$= \begin{cases} \dfrac{1}{300000}(x - 2500000), & 0 \le x \le 2800000 \\ \dfrac{1}{200000}(3000000 - x), & x \ge 2800000 \end{cases}.$$

When the supplied fund $x = 1,200,000$ yuan,

$$K_0(P) = k(1200000) = \frac{1}{300000}(1200000 - 2500000) = -\frac{13}{3} < 0.$$

Namely, for E, the problem $P = G * L$ is an incompatible problem.

Example 7.3.2

In the Three Kingdoms' time period, someone gave Caocao an elephant as a gift. Caocao wondered about the elephant's weight. However, the elephant's weight was far beyond the maximum measurement capability of the scales used in that era, and the elephant must be kept alive. None of Caocao's secretaries had a feasible strategy. Apparently, this is an incompatible problem. Establish the problem's extension model and compatible degree function, and judge its incompatible degree.

Solution: The problem's extension model is

$$P = G * L$$

$$
= \begin{bmatrix} \text{Weigh,} & \text{dominating object,} & \begin{bmatrix} \text{Elephant } D_1, & \text{weight,} & x_{01} \text{ kg} \\ & \text{decomposability,} & x_{02} \end{bmatrix} \\ & \text{acting object,} & \text{Caocao's secretaries} \end{bmatrix}
$$

$$
* \begin{bmatrix} \text{Tools,} & \text{type,} & (\text{Steelyard } D_2, \text{measure, } [0, 200] \text{ kg}) \\ & \text{time,} & \text{Three Kingdoms period} \end{bmatrix}
$$

and $x_{01} \gg 200$, x_{02} = cannot be divided into several parts that weight less than 200 kg. c_{01s} = weight; c_{02s} = decomposability; c_{01s}, c_{02s} are the characteristics of the elephant D_1 required by the condition L; X_{01} = [0, 200] and X_{02} = $\{x'_{02}\}$. Where x'_{02} = can be divided into several parts that weight less than 200 kg. Suppose c_{0t} = "measuring capacity," which is the supplied characteristic by the condition.

$$
g_0 = \begin{bmatrix} D_1, & c_{01s}, & x_{01} \text{ kg} \\ & c_{02s}, & x_{02} \end{bmatrix},
$$

$$
l_0 = \left(D_2, \quad c_{0t}, \quad [0, \quad 200] \text{kg} \right).
$$

The extension model of the problem P's core problem is

$$
P_0 = g_0 * l_0 = \begin{bmatrix} D_1, & c_{01s}, & x_{01} \text{ kg} \\ & c_{02s}, & x_{02} \end{bmatrix} * \left(D_2, \quad c_{0t}, \quad [0, 200] \text{kg} \right).
$$

Respectively taking X_{01} and X_{02} as the positive fields, establish the problem's compatible degree function as

$$
y = k_1(x_1) \wedge k_2(x_2)
$$

where

$$k_1(x_1) = \frac{200 - x_1}{200},$$

$$k_2(x_2) = \begin{cases} -1, & x_2 = x_{02} \\ 1, & x_2 = x'_{02} \end{cases}$$

Obviously, problem P's compatible degree is:

$$K_0(P) = k_1(x_{01}) \wedge k_2(x_{02}) = \frac{200 - x_{01}}{200} \wedge (-1) < 0, \text{ when } x_{01} \gg 200.$$

So problem P is an incompatible problem. In other words, for Caocao's secretaries, the problem is an incompatible problem.

7.3.2 The Basic Procedure and Flowchart of the ECIG Method for Solving Incompatible Problems

The basic procedure of the extension creative idea generation (ECIG) method for solving incompatible problems consists of the following steps:

1. Define the problem's goal and condition, and establish the problem's extension model by using basic-element formalized expressing system.
2. Study the goal and condition to see if they are variable; select an evaluation characteristic; determine the problem's core problem, according to the values (or value range) supplied by the problem and the values (or value range) required by achieving the goal.
3. Establish the incompatible problem's compatible degree function by using a dependence function, and compute the problem's incompatible degree.
4. Determine the order of study, i.e. the goal first or the condition first:

    a. If the goal stays constant, analyze the condition first; adopt the correlative analysis method to establish the problem's correlative tree (network).
    b. If the condition stays constant, analyze the goal first; adopt the implicating analysis method to establish the problem's implicating tree.
    c. If the goal and condition are both variable, execute the two previous steps in order, and then establish the problem's correlative-implicating tree.

5. Perform divergence analysis or conjugated analysis on the leaves of the correlative tree or implicating tree; then perform an extension transformation and a conductive transformation implicating tree according to the conductive transformation theory; the tree formed by extension transformation and conductive transformation is usually called an extension creative idea generation tree.

6. For the problem formed after the transformation, compute the value of its compatible function; if the compatible degree changes from being less than 0 or equal to 0 to being larger than 0 (For some practical problems, being equal to 0 is regarded as compatible; you should perform a definite analysis according to the given problems), the transformation or transformation operation is regarded as the extension creative idea of solving the incompatible problem.

7. For multiple extension creative ideas generated after applying the transformation, select measuring indicators according to the practical problem's requirements, and then evaluate the creative ideas using the superiority evaluation method and sort them in descending order according to their superiorities.

The flowchart of this procedure is given in Figure 7.3.1.

The following presents a few case studies to demonstrate how the method of solving incompatible problems by transforming either the condition or goal is practically employed.

Case Analysis:

Example 7.3.3

In Example 7.3.1, we established the problem's extension model and judged it to be an incompatible problem. Provide an extension creative idea to solve this incompatible problem.

Solution: Because the new apartment W must be bought, the condition l_0 needs to be transformed to solve the problem. The condition is a real resource one, so the conjugated analysis should be performed on person E's resources to find a superior resource.

According to the method of divergent analysis, we have

$$
l_0 \dashv
\begin{cases}
\left(\text{Bank } O_1, \quad c_{0t}, \quad v_1 \text{ yuan} \right) \\[4pt]
\left(\text{Parents } O_2, \quad c_{0t}, \quad v_2 \text{ yuan} \right) \\[4pt]
\left(\text{Friends } O_3, \quad c_{0t}, \quad v_3 \text{ yuan} \right) \\[4pt]
\left(\text{Network financial platform } O_4, \quad c_{0t}, \quad v_4 \text{ yuan} \right)
\end{cases}
\quad .
$$

Through a resource analysis, person E's superior resource is the virtual resource—reputation resource and soft resource—and relation resource: person E has a good reputation with the banking system and the network financial platform, which can help him obtain loans with relatively low interest rates. Moreover, person E has a very good relationship with his parents and friends.

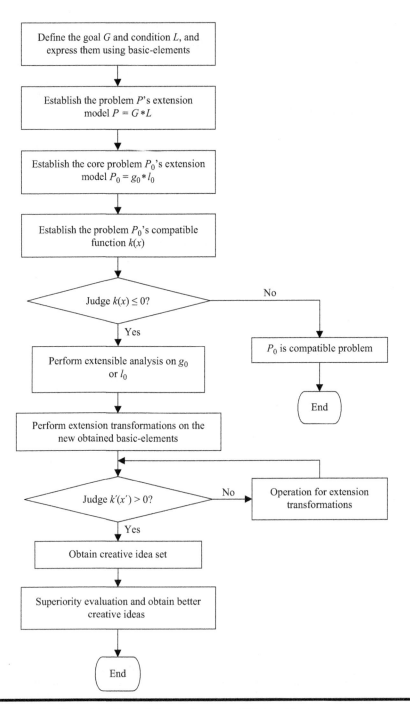

Figure 7.3.1   Flowchart for solving an incompatible problem.

Therefore, we can perform at least the following four transformations on the condition:

1. $T_1 l_0 = (E \oplus O_1, c_{0t}, 1200000 \oplus 1600000 \text{ yuan}) = l_1', k(x_1') = k(2800000) = 1 > 0.$
   That is, T_1 is to apply for a 1,600,000 yuan loan from a bank and then combine the loan with his own fund.

2. $T_2 l_0 = (E \oplus O_2, c_{0t}, 1200000 \oplus 600000 \text{ yuan}) = l_2', k(x_2') = k(1800000) = -\frac{7}{3} < 0.$
   That is, T_2 is to borrow 600,000 yuan from his parents and combine the money with his own fund.

3. $T_3 l_0 = (E \oplus O_3, c_{0t}, 1200000 \oplus 1000000 \text{ yuan}) = l_3', k(x_3') = k(2200000) = -1 < 0.$
   That is, T_3 is to borrow 1,000,000 yuan from friends and combine the money with his own fund.

4. $T_4 l_0 = (E \oplus O_4, c_{0t}, 1200000 \oplus 1500000 \text{ yuan}) = l_3', k(x_4') = k(2700000) = \frac{2}{3} > 0.$
   That is, T_4 is to apply for loans for the amount of 1,500,000 yuan from the network financial platform and combine the loans with his own fund.

Among these four transformations, only T_1 and T_4 can make $k(x_i') \geq 0, i = 1, 4$. Namely, both T_1 and T_4 can help solve the incompatible problem. Although T_2 and T_3 cannot help solve the incompatible problem, we have

$$T_5 l_0 = (E \oplus O_2 \oplus O_3, c_{0t}, 2800000 \text{ yuan}) = l_5', k(x_5') = k(2800000) = 1 > 0.$$

Obviously, $T_5 = T_2 \wedge T_3$ can also solve the problem. T_5 means simultaneously apply both T_2 and T_3 (their "and") to solve the problem.

Transformations T_1, T_4, and T_5 form three different creative ideas that can be used to solve the incompatible problem. Next, perform a superiority evaluation on the obtained creative ideas according to person E's particular conditions, such as the bank rate, loan repayment ability (income), repayment period, etc.

Example 7.3.4

In Example 7.3.2, we judged problem P to be an incompatible problem. In other words, for Caocao's secretaries, the problem is incompatible.

Solution: Evidently the problem cannot be solved by directly transforming the given condition. Namely, in that era, no scale was large enough to weigh the elephant. Thus, we must consider transforming the goal, satisfying the condition that if the new goal is achieved, the original goal is also achieved.

$$g = \begin{bmatrix} O, & c_{01s}, & x_{01} \text{ kg} \\ & c_{02s}, & x_{02}' \end{bmatrix}, \text{ and } g \Rightarrow g_0$$

For the problem, according to the divergent tree method (one characteristic multiple objects), we have

$$g_0 \dashv \left\{ g_1 = \begin{bmatrix} \text{Rocks } S_1, & c_{01}, & x_{01} \text{ kg} \\ & c_{02}, & x_{02}' \end{bmatrix}, \quad g_2 = \begin{bmatrix} \text{Sandpile } S_2, & c_{01}, & x_{01} \text{ kg} \\ & c_{02}, & x_{02}' \end{bmatrix}, \right.$$

$$\left. g_3 = \begin{bmatrix} \text{Woods } S_3, & c_{01}, & x_{01} \text{ kg} \\ & c_{02}, & x_{02}' \end{bmatrix}, \dots \right\}$$

According to the decomposability, $g_i / \{g_{i1}, g_{i2}, \ldots, g_{in_i}\}$, $i = 1, 2, 3, \cdots$, where

$$g_{ij} = \begin{bmatrix} S_{ij}, & c_{01}, & x_{ij} \text{ kg} \\ & c_{02}, & x'_{02} \end{bmatrix}, \quad x_{ij} < 200 \text{ kg}, \quad j = 1, 2, \ldots, n_i.$$

Then, for convenience, select one g_i as the new goal implicating g_0.

Now, we must determine whether "g_i is equivalent to g_0" in the situation that x is unknown and cannot be measured using any of the available scales. The goal G "weigh the elephant" has changed to be goal G_1 "find a weighing apparatus that can measure 'g_i is equivalent to g_0'" (Note: This is the goal of Caocao's son, Caochong.). Apparently

$$G \Leftarrow G_1, G_1 = \left(\text{Measure}, \quad \text{dominating object}, \quad g_0 \Leftrightarrow g_i \right)$$

According to the method of correlative analysis: On land we have

$$\left(O, \quad \text{weight}, \quad x_{01} \right) \sim \left(O, \quad \text{volume}, \quad z \right), z = f(x_{01}).$$

In the water, we have

$$\left(O, \quad \text{weight}, \quad x_{01} \right) \sim \left(O, \quad \text{buoyancy}, \quad z_1 \right) \sim \left(\text{Carrier}, \quad \text{draft (of water)}, \quad z_2 \right).$$

By performing an implicating analysis according to what we know, we have

$$G_1 \Leftarrow G_{11} \vee G_{12}$$

where

$$G_{11} = \begin{bmatrix} \text{Measure}, & \text{dominating object}, & g_0 \Leftrightarrow g_i \\ & \text{tools}, & \text{lever} \vee \text{crotch} \vee \cdots \\ & \text{location}, & \text{land} \end{bmatrix}$$

$$G_{12} = \begin{bmatrix} \text{Measure}, & \text{dominating object}, & g_0 \Leftrightarrow g_i \\ & \text{tools}, & \text{ship} \vee \text{wood raft} \vee \cdots \\ & \text{location}, & \text{water} \end{bmatrix}$$

The weighing apparatus that can determine whether "g_i is equivalent to g_0" involves lever, ship, and so on. If we utilize the lever on land, a huge container is required to contain the elephant and some stuff with an equivalent weight, which is hard to practically implement. So, we utilize a ship in water. According to the knowledge that a same depth of immersion always indicates a same weight of the items on the ship, we can put the elephant on the deck of the ship, mark the depth of immersion on the ship, and then take the elephant off the ship and put other stuff on ship until the marked depth of immersion is reached. In this way, the goal G_{12} can be achieved, and the goal G_1 is also achieved. Obviously, the ship is the most convenient weighing apparatus.

Perform first the substituting transformation $T_i g_0 = g_i$, and then the decomposing transformation $T_i' g_i = \{g_{i1}, \ g_{i2}, \ \cdots, \ g_{in_i}\}$, where

$$
g_{ij} = \left[\begin{array}{ccc} S_{ij}, & c_{01}, & x_{ij}\ \text{kg} \\ & c_{02}, & x'_{02} \end{array} \right]
$$

and $x_{i1} + x_{i2} + \cdots + x_{in_i} = x_{01}$, $x_{ij} < 200$, $(i = 1, 2, 3, \cdots, j = 1, 2, \cdots, n_i)$. Then we have

$$
K(T_i' T_i g_0) = \bigwedge_{j=1}^{n_i} k(T_i' g_i) = \bigwedge_{j=1}^{n_i} [k_1(x_{ij}) \wedge k_2(x'_{02})] = \bigwedge_{j=1}^{n_i} \left(\frac{200 - x_{ij}}{200} \wedge 1 \right) > 0.
$$

Namely, perform the transformation $T_i' T_i = \{$First, we use a ship to measure some decomposable stuff S_i having equivalent weight with the elephant; then substitute S_i for the elephant; and decompose S_i into weighable stuff $S_{ij}\}$, which makes the incompatible problem compatible.

This was Caochong's method. He utilized the theory that the elephant and the stones have the same depth of immersion to find an equivalent goal so that only a ship and an available scale were used for weighing the elephant.

In fact, Caochong's method is not the smartest. If some men and stones are both used, we could achieve the matching depth of immersion more quickly, right?

Thinking: If there is no water, no ship, and no platform scale, how can you then weigh the elephant?

7.4 Converting Bridge Method for Solving Antithetical Problems

QUESTIONS AND THINKING:

- Hong Kong's traffic drives on the left side of the road, while Shenzhen's traffic drives on the right side of the road. How can one connect these two places by using a bridge without causing traffic collision?
- Two individuals eat a hotpot dinner together. One likes spicy soup, while the other likes non-spicy soup. How can they select the soup type?
- In these problems, two antithetical goals should be implemented under the current condition. What is the solution? Is there any law to obey to solve these problems?

For an antithetical problem, the following three types of methods are generally adopted.

1. "Either this or that" fighting method, e.g. "approve one, deny the other," "my way or your way." Such a method is simple, direct, but is easy to bring new contradiction.
2. The balancing method of "both one and the other," also called the compromising method, e.g. "you get 30 percent and I get 70 percent." Both sides share the profit by bargaining so that, at the compromise point, the sides reconcile any and all contradictions, which is usually implemented through negotiations.
3. The converting bridge method of "each goes his own way": This is an ingenious method to solve the contradictory problem. In the following section, we will introduce how to solve the antithetical problem by using such method.

Each antithetical problem is also a common type of contradictory problem. In this section, based on the extension model introduced in Section 7.2, we first present the antithetical problem's extension model, and then give the converting bridge method of solving antithetical problems.

7.4.1 An Antithetical Problem's Extension Model

Suppose you have a problem, $P = (G_1 \wedge G_2) * L$, where G_1, G_2, and L are basic-elements or compounded-elements. If the goal G_1 and G_2 cannot be achieved simultaneously, the problem is called an antithetical problem, denoted as $(G_1 \wedge G_2) \uparrow L$.

Given an antithetical problem $P = (G_1 \wedge G_2) * L, (G_1 \wedge G_2) \uparrow L$, if there is a transformation $T = (T_{G_1}, T_{G_2}, T_L)$ that makes

$$(T_{G_1} G_1 \wedge T_{G_2} G_2) \downarrow T_L L,$$

then the transformation T is called a solution transformation of problem P, which makes G_1 and G_2 coexist.

The transformed object of the solution transformation is essential to converting antithesis into coexistence. Because it plays the role of a converting media, it is vividly called a converting bridge, denoted as $Br(G_1, G_2)$.

Reference **Extenics** gives an extension set-based definition of an antithetical problem, establishing a method of an extension model and a corresponding method of forming judgments. Due to the relative complexity of this, the method of establishing the coexisting degree function of an antithetical problem is not presented here. After confirming the given problem as antithetical, we discuss a method to solve it by using the converting bridge method, of which the core is "each goes his own way."

7.4.2 Converting Bridge Method

In our previous discussions, we considered two types of converting bridges: the connecting converting bridge and the separating converting bridge. Through recent research in practice, we discovered that neither the connecting converting bridge nor the separating converting bridge is isolated. Some connecting converting bridges have separating functions, while some separating converting bridges have connecting functions. For example, an overpass of the highway system is not only a connection, but it also separates the vehicles in vertical and horizontal directions. Likewise, bunk beds are designed to solve the problem that the available space cannot contain two beds. The bunk bed structure not only separates the two beds but also connects them.

In terms of their function (or role), a converting bridge includes connecting-oriented "connecting–separating" converting bridges as well as separating-oriented "separating-connecting" converting bridges, where the "connecting-separating" part or "separating-connecting" part is called the transition part. In other words, a converting bridge is used to construct a transition part to connect or separate both sides so that their coexistence can be achieved. In the following section, we present some examples to demonstrate how to construct the transition parts of the two types of converting bridges. For the corresponding formalized methods, refer to the monograph ***Extenics***.

Case Analysis:

Example 7.4.1

Traffic in Hong Kong drives on the left side of the road, while vehicles in neighboring Shenzhen drive on the right side of the road. How can one connect these two places? If they are connected directly, there will be traffic collisions in the conjunction. The Huanggang bridge of Shenzhen is a connecting-separating transition part between these two traffic systems. On the bridge, vehicles going from Hong Kong to Shenzhen shift to the right side of the road after entering Shenzhen, while vehicles going from Shenzhen to Hong Kong automatically shift to the left side of the road after entering Hong Kong. Thus, "each goes his own way" can be practically implemented.

Example 7.4.2

A couple eats a hotpot dinner together (at the same time t). One likes spicy soup, while the other doesn't. How can they select the soup type to meet the preferences of both diners? This is an antithetical problem, which has many solutions. If non-spicy soup is selected, the one liking spicy soup can eat with spicy seasoning. If individual hotpots are adopted, the problem is also solved. If only one hotpot is used, a separating board can be added into the pot, dividing the pot into two individual parts, in which spicy soup and non-spicy soup can be respectively adopted. This is the idea of "Yuanyang hotpot."

Example 7.4.3

Put a wolf and a chicken in the same cage and ensure that the chicken is not eaten by the wolf. This is a problem that is simple to state, but it is also an antithetical problem. The same method as used in the above example can be adopted. We can put a separating board in the cage as the separating-connecting transition part to solve the antithetical problem.

Example 7.4.4

A land agent plans to build a business building in a place that is h square meters. Considering the potential commercial profits, the larger the area covered by the building, the better. According to the city design rules, a 30% green coverage ratio should be ensured. However, the designed building's construction area is 90% of h square meters, leaving at most 10% green coverage ratio. Apparently, this is a contradictory problem.

According to regular thinking, the design scheme of the building must be modified. The building's construction area should be shrunk, namely altering the goal G_1. If this is done, however, the sales revenue will have to decrease. The condition L can also be altered, namely additionally buying some land as the green coverage area. This, however, will increase the cost.

Using the converting bridge method, the "top" of the building can be a separating-connecting transition part between the land of the business building and the green land, forming an "air garden." Thus the problem can be solved.

In various fields of human endeavors, there are many antithetical problems. For example, a method of traditional competitions among different enterprises is to defeat the opponents. The relationship among the opponents is antithetical. With society's development, the business community gradually realized that "win–win" situations or even "multiple-win" situations are better ways of survival and growth. Only collaboration can avoid unnecessary losses. The converting bridge method can be applied to solving antithetical problems in different endeavors with excellent generality.

Thinking and Exercises

1. Using the definition of extension sets, taking consumers' purchase ability and purchase willingness as evaluating characteristics, and taking a promoting sale activity as the extension transformation, perform a variable categorization on the enterprise's market.
2. Using the definition of extension sets, taking students' organizing ability, linguistic expressing ability, and computer skills as evaluation characteristics, and taking trainings corresponding to these three abilities as extension transformations, perform a variable categorization on the students.
3. On your camping trip, you plan to boil a pot of water, but there isn't enough firewood. What should you do? What are the problem's goals and conditions? How do you establish the extension model? Establish the problem's compatible degree function and compute the problem's compatible degree.

4. You plan to convert a colorful four-page document into an electronic JPG file with a size less than 200 kb with the contents being clear. How do you do it? What are the problem's goals and conditions? How do you establish an extension model? Establish the problem's compatible degree function and compute the problem's compatible degree.

5. Someone plans to build a golf driving range and a fishing pool on a same parcel of land. How do you implement this plan? What are the problem's goals and conditions? Is it an incompatible problem or an antithetical problem? How do you solve it?

6. On the balcony, there is a 20-cm deep washing basin with a tap that is 70 cm away from the ground. If someone wants to fill up a 50 cm tall bucket with water, do you have a feasible approach to resolve this problem? Use the extension creative idea generating method to generate multiple solutions and point out the superior ones.

7. You want to move a 100-kg security box from the living room into the bedroom. There is no other person to help you, and the finished floor cannot be damaged. How do you do it? Use the extension creative idea generating method to generate multiple solutions and list the superior ones.

8. You want to buy a 4,000-yuan notebook computer and a 3,000-yuan mobile phone, but you have only 4,000 yuan. How can you make the purchases? Provide your extension creative ideas by using the converting bridge method.

9. The queen bee is the only female bee that can normally spawn in the bee population. If a larva is arranged to live in the queen's cell to eat royal jelly when it is young, it then later becomes the queen bee. A fertilized spawn of the queen bee becomes a worker bee, while non-fertilized spawn of the queen bee becomes a male bee. Worker bees occupy a dominant percentage in the bee population. Worker bees are female bees in which the reproductive organ is abortive. They have no reproductive ability and are relatively small in physical size. They are responsible for collecting flower pollen, making honey, and

feeding larva and the queen bee. In a bee box, the more worker bees there are, the greater the output of honey. However, the queen bee's reproductive ability is limited. How can one enhance the output of honey? Someone proposes putting two queen bees in a same bee box. Theoretically, if there are two queen bees in a bee box, the number of worker bees will be doubled and the output of honey can also be potentially doubled. Queen bees, however, can fight with each other. This is an antithetical problem. How do you solve this problem?

Chapter 8

Extension Creative Idea Generating Methods for Products

Content Summary

Product innovation is an eternal theme for enterprises. Enterprises can grasp the market condition correctly if they can grasp laws and methods of product innovation. The product extension creative idea generating method is to study the rules and methods of generating creative product ideas on the basis of the possibility and orderliness of formalized researches, extensions and transformations on customers' needs, products and their functions, etc., so as to generate new product ideas in batches, develop new product development software and predict the appearance of new products.

In this chapter, the convenient and simple extension innovation four-step method is presented before the introduction of three creation methods for product innovation.

8.1 Extension Innovation Four-Step Method

QUESTIONS AND THINKING:

- Where do new product ideas come from?
- Are there any laws or methods that can be followed to create new products?
- Are there any convenient and easy-to-learn methods that can help product innovators create new product ideas?

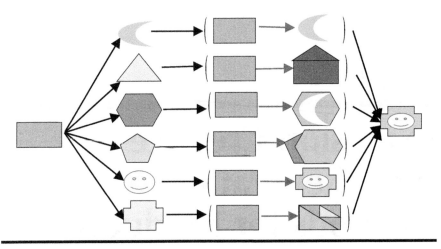

Figure 8.1.1 Diagram of the extension innovation four-step method.

The extension innovation four-step method is a universal method that utilizes the basic extension method to innovate or to solve contradictory problems. This method consists of four steps: modeling, extension, transformation, and selection. By following these four steps, new product ideas or ideas for solving contradictory problems can be generated.

The extension innovation four-step method is presented in Figure 8.1.1.

By taking product innovation as an example, the steps of the extension innovation four-step method can be outlined as follows:

1. **Modeling:** The extension model construction method introduced in Chapter 2 is utilized to set up an extension model for the product by means of formalization based on the specific problem given. The extension model includes a matter-element model of the product, a functional affair-element model, a demand affair-element model, a structure relation-element model, and a compound element model for complex products, which can be regarded as a starting point for innovation.

2. **Extension:** To find multiple approaches to innovate, the extensible analysis method and the conjugate analysis method introduced in Chapters 3 and 4 are utilized to extend the product's extension model established in step 1, which can be a basis for generating creative ideas.

3. **Transformation:** By using the extension transformation method introduced in Chapter 5 as a tool for generating ideas, the original product's extension model can be transformed into all kinds of extended models to obtain a variety of creative ideas of new products.

4. **Selection:** The new product ideas generated in the previous step are evaluated by using the superiority evaluation method introduced in Chapter 6 so that ideas with higher superiority can be selected as alternative ideas of new products.

Case Analysis:

Example 8.1.1

Assume that an existing ceramic cup D is composed of the cup body D_1, the cup handle D_2, the cup lid D_3, and the cup bottom D_4. Let's generate a new product idea by using the extension innovation four-step method.

Step 1 – Modeling: Extension models of the four components of cup D and their relations are established. For the purpose of illustrating the operational steps of the method, only the shape and color of the cup are considered in the compositions of the cup. Likewise, only the connection relationship O_1 between the cup body and the cup handle as well as the up and down relationship O_2 between the cup lid and the cup body are considered in the relationships among the components. The modeling is not conducted for the functions of the cup.

$$
M_1 = \begin{bmatrix} D_1, & \text{shape,} & \text{cylinder} \\ & \text{color,} & \text{white} \end{bmatrix}, M_2 = \begin{bmatrix} D_2, & \text{shape,} & \text{semicircle ring-like} \\ & \text{color,} & \text{white} \end{bmatrix},
$$

$$
M_3 = \begin{bmatrix} D_3, & \text{shape,} & \text{spherical crown} \\ & \text{color,} & \text{white} \end{bmatrix}, M_4 = \begin{bmatrix} D_4, & \text{shape,} & \text{circular} \\ & \text{color,} & \text{white} \end{bmatrix},
$$

$$
R_1 = \begin{bmatrix} O_1, & \text{antecedent,} & \text{cup body } D_1 \\ & \text{consequent,} & \text{cup handle } D_2 \\ & \text{means,} & \text{an organic whole} \end{bmatrix},
$$

$$
R_2 = \begin{bmatrix} O_2, & \text{antecedent,} & \text{cup lid } D_3 \\ & \text{consequent,} & \text{cup body } D_1 \\ & \text{means,} & \text{directly fastened} \end{bmatrix}.
$$

Step 2 – Extension: A divergence' analysis is conducted respectively on the above-mentioned product matter-element models and the structure relation-element model by using the divergent tree method in the extensible analysis method as an example. Moreover, a correlation analysis, implicit analysis, and extensible analysis can be conducted on those models, too, all of which will be omitted here. In this chapter, only the divergence rule of "a characteristic with multiple values" is utilized. The divergence result is presented as follows:

$$M_1 = \begin{bmatrix} D_1, & \text{shape,} & \text{cylinder} \\ & \text{color,} & \text{white} \end{bmatrix} \dashv \begin{cases} M_{11} = \begin{bmatrix} D_{11}, & \text{shape,} & \text{cartoon images} \\ & \text{color,} & \text{color of cartoon images} \end{bmatrix} \\[2ex] M_{12} = \begin{bmatrix} D_{12}, & \text{shape,} & \text{animal character} \\ & \text{color,} & \text{color of animal character} \end{bmatrix} \\[2ex] M_{13} = \begin{bmatrix} D_{13}, & \text{shape,} & \text{cylinder with heart shape} \\ & \text{color,} & \text{white} \end{bmatrix} \\[2ex] M_{14} = \begin{bmatrix} D_{14}, & \text{shape,} & \text{quadrangular} \\ & \text{color,} & \text{red} \end{bmatrix} \end{cases}$$

$$M_2 = \begin{bmatrix} D_2, & \text{shape,} & \text{semicircle ring-like} \\ & \text{color,} & \text{white} \end{bmatrix} \dashv \begin{cases} M_{21} = \begin{bmatrix} D_{21}, & \text{shape,} & \text{ear-like} \\ & \text{color,} & \text{color of some cartoon image} \end{bmatrix} \\[2ex] M_{22} = \begin{bmatrix} D_{22}, & \text{shape,} & \text{tail-like} \\ & \text{color,} & \text{color of some animal's tail} \end{bmatrix} \\[2ex] M_{23} = \begin{bmatrix} D_{23}, & \text{shape,} & \text{torus} \\ & \text{color,} & \text{red} \end{bmatrix} \\[2ex] M_{24} = \begin{bmatrix} D_{24}, & \text{shape,} & \text{rectangular annulus} \\ & \text{color,} & \text{black} \end{bmatrix} \end{cases}$$

$$M_3 = \begin{bmatrix} D_3, & \text{shape,} & \text{spherical crown} \\ & \text{color,} & \text{white} \end{bmatrix} \dashv \begin{cases} M_{31} = \begin{bmatrix} D_{31}, & \text{shape,} & \text{head of some cartoon image} \\ & \text{color,} & \text{color of some cartoon image's head} \end{bmatrix} \\[2ex] M_{32} = \begin{bmatrix} D_{32}, & \text{shape,} & \text{head of some animal} \\ & \text{color,} & \text{color of some animal's head} \end{bmatrix} \\[2ex] M_{33} = \begin{bmatrix} D_{33}, & \text{shape,} & \text{disc} \\ & \text{color,} & \text{red} \end{bmatrix} \\[2ex] M_{34} = \begin{bmatrix} D_{34}, & \text{shape,} & \text{rectangular} \\ & \text{color,} & \text{black} \end{bmatrix} \end{cases}$$

$$M_4 = \begin{bmatrix} D_4, & \text{shape,} & \text{circular} \\ & \text{color,} & \text{white} \end{bmatrix} \dashv \begin{cases} M_{41} = \begin{bmatrix} D_{41}, & \text{shape,} & \text{heart shape} \\ & \text{color,} & \text{red} \end{bmatrix} \\ M_{42} = \begin{bmatrix} D_{42}, & \text{shape,} & \text{head of some animal} \\ & \text{color,} & \text{color of some animal's head} \end{bmatrix} \\ M_{43} = \begin{bmatrix} D_{43}, & \text{shape,} & \text{head of some cartoon image} \\ & \text{color,} & \text{color of some cartoon image's head} \end{bmatrix} \\ M_{44} = \begin{bmatrix} D_{44}, & \text{shape,} & \text{rectangular} \\ & \text{color,} & \text{black} \end{bmatrix} \end{cases}$$

$$R_1 = \begin{bmatrix} O_1, & \text{antecedent,} & D_1 \\ & \text{consequent,} & D_2 \\ & \text{means,} & \text{an organic whole} \end{bmatrix}$$

$$\dashv \begin{cases} R_{11} = \begin{bmatrix} O_{11}, & \text{antecedent,} & D_{1i} \\ & \text{consequent,} & D_{2j} \\ & \text{means,} & \text{separated} \end{bmatrix} \\ R_{12} = \begin{bmatrix} O_{12}, & \text{antecedent,} & D_{1i} \\ & \text{consequent,} & D_{2j} \\ & \text{means,} & \text{none} \end{bmatrix} \\ R_{13} = \begin{bmatrix} O_{13}, & \text{antecedent,} & D_{1i} \\ & \text{consequent,} & D_{2j} \\ & \text{means,} & \text{an organic whole in the top with separated in the bottom} \end{bmatrix} \end{cases}$$

$$R_2 = \begin{bmatrix} O_2, & \text{antecedent,} & D_3 \\ & \text{consequent,} & D_1 \\ & \text{means,} & \text{directly fastened} \end{bmatrix}$$

$$\dashv \left\{ \begin{array}{l} R_{21} = \begin{bmatrix} O_{21}, & \text{antecedent,} & D_{1i} \\ & \text{consequent,} & D_{3t} \\ & \text{means,} & \text{directly fastened} \end{bmatrix} \\ \\ R_{22} = \begin{bmatrix} O_{22}, & \text{antecedent,} & D_{3t} \\ & \text{consequent,} & D_{1i} \\ & \text{means,} & \text{screw type} \end{bmatrix} \\ \\ R_{23} = \begin{bmatrix} O_{23}, & \text{antecedent,} & D_{1i} \\ & \text{consequent,} & D_{3t} \\ & \text{means,} & \text{embedded type} \end{bmatrix} \end{array} \right.$$

$$(i = 1, 2, 3, 4; \; j = 1, 2, 3, 4; \; t = 1, 2, 3, 4).$$

Step 3 – Transformation: Taking a substituting transformation and the AND operation as examples, many new product models can be formed. First, substituting transformations are conducted as follows:

$$T_{12}M_1 = M_{12} = \begin{bmatrix} D_{12}, & \text{shape,} & \text{some animal character} \\ & \text{color,} & \text{color of some animal character} \end{bmatrix}$$

$$T_{13}M_1 = M_{13} = \begin{bmatrix} D_{13}, & \text{shape,} & \text{cylinder with heart shape} \\ & \text{color,} & \text{white} \end{bmatrix}$$

$$T_{22}M_2 = M_{22} = \begin{bmatrix} D_{22}, & \text{shape,} & \text{tail-like} \\ & \text{color,} & \text{color of some animal's tail} \end{bmatrix}$$

$$T_{24}M_2 = M_{24} = \begin{bmatrix} D_{24}, & \text{shape,} & \text{rectangular annulus} \\ & \text{color,} & \text{black} \end{bmatrix}$$

$$T_{33}M_3 = M_{33} = \begin{bmatrix} D_{33}, & \text{shape,} & \text{disc} \\ & \text{color,} & \text{red} \end{bmatrix}$$

$$T_{41}M_4 = M_{41} = \begin{bmatrix} D_{41}, & \text{shape,} & \text{heart shape} \\ & \text{color,} & \text{red} \end{bmatrix}$$

$$T_{42}M_4 = M_{42} = \begin{bmatrix} D_{42}, & \text{shape,} & \text{head of some animal} \\ & \text{color,} & \text{color of some animal's head} \end{bmatrix}$$

$$T_{R21}R_2 = R_{21} = \begin{bmatrix} O_{21}, & \text{antecedent,} & D_{12} \\ & \text{consequent,} & D_{33} \\ & \text{means,} & \text{directly fastened} \end{bmatrix}$$

And then, composite transformations are conducted as follows:

1. $T = T_{R21}\left(T_{12} \wedge T_{22} \wedge T_{33} \wedge T_{42}\right)$, thus

$S = M_{12} \wedge M_{22} \wedge M_{33} \wedge M_{42} \wedge R_{21}$

$$= \begin{bmatrix} D_{12} \wedge D_{22} \wedge D_{33} \wedge D_{42}, & \text{shape,} & \text{some animal's body} \wedge \text{some animal's tail} \\ & & \wedge \text{disc} \wedge \text{some animal's head} \\ & \text{color,} & \text{some animal's color} \end{bmatrix}.$$

$$\wedge \begin{bmatrix} O_{21}, & \text{antecedent,} & D_{12} \\ & \text{consequent,} & D_{33} \\ & \text{means,} & \text{directly fastened} \end{bmatrix}$$

2. $T' = T_{13} \wedge T_{24} \wedge T_{41}$, thus

$S' = M_{13} \wedge M_{24} \wedge M_{41}$

$$= \begin{bmatrix} D_{13} \wedge D_{24} \wedge D_{41}, & \text{shape,} & \text{cylinder with heart shape} \wedge \text{rectangular annulus} \\ & & \wedge \text{heart shape} \\ & \text{color,} & \text{white} \wedge \text{black} \wedge \text{red} \end{bmatrix}.$$

Models S and S' obtained through the above two transformations are two new product models, the corresponding new product ideas of which can be presented in words as follows:

Creative idea S: The cup's body is in a shape of some animal body. The cup's handle is in the shape of the animal's tail. The cup's lid is disc-shaped, and the cup's bottom is in the shape of the animal head. Moreover, it's a covered cup with the cup's body on top and the cup's lid on bottom.

Creative idea S': It is an uncovered cup with a heart-shaped cylindrical body, a rectangular handle, and a heart-shaped bottom.

Step 4 – Selection: By using the superiority evaluation method with novelty and cost as evaluating characteristics, and (novelty, high) and (cost, [50,100] yuan) as measuring indicators, we can compute that the superiority of the cost is 80 yuan. The divergent correlation function and the simple correlation function are presented respectively as follows:

$$k_1(x_1) = \begin{cases} 1, & x_1 = \text{high} \\ 0, & x_1 = \text{medium} \\ -1, & x_1 = \text{low} \end{cases}$$

$$k_2(x_2) = \begin{cases} \dfrac{x_2 - 50}{80 - 50} = \dfrac{x_2 - 50}{30}, & x_2 \le 80 \\ \dfrac{100 - x_2}{100 - 80} = \dfrac{100 - x_2}{20}, & x_2 \ge 80 \end{cases}.$$

The novelty and cost of creative idea S are $x_{11} = \text{high}$ and $x_{21} = 90$ yuan. Thus, its dependence degree is $k_1(x_{11}) = 1$, $k_2(x_{21}) = k_2(90) = 0.5$.

The novelty and cost of creative idea S' are $x_{12} = \text{medium}$ and $x_{22} = 70$ yuan. Thus, its dependence degree is $k_1(x_{12}) = 0$, $k_2(x_{22}) = k_2(70) = 0.67$.

Weights of these two measuring indicators can be selected as $\alpha_1 = 0.8, \alpha_2 = 0.2$. So the comprehensive superiority of creative idea S is

$$C(S) = \alpha_1 k_1(x_{11}) + \alpha_2 k_2(x_{21}) = 0.8 + 0.2 * 0.5 = 0.9$$

and the comprehensive superior of creative idea S' is

$$C(S') = \alpha_1 k_1(x_{12}) + \alpha_2 k_2(x_{22}) = 0 + 0.2 * 0.67 = 0.134.$$

The comprehensive superiority of the creative idea S is higher. Therefore, S is selected as a new product idea to develop.

Illustration: The example shows the operability of the extension innovation four-step method. Various product innovation methods mentioned later in this chapter are derived on the basis of this method with different starting points.

8.2 The First Creation Method: Generating Creative Ideas of Products Based on Customers' Needs

QUESTIONS AND THINKING:

- A person who will go on a business trip or a journey would like to bring only a suitcase or a traveling bag with several pairs of shoes for adapting to different occasions, like leather shoes, cloth shoes, climbing boots, running shoes, etc. Can you help her make her decision on shoes?
- What needs do consumers have? How can we obtain creative ideas of new products through associating consumers' needs with product functions?

Studying the laws that govern product innovation must start from studying the needs of consumers because consumers buy products to meet their own needs, rather than for the sake of the product itself.

The company that identifies needs that are unmet, i.e., needs that can be improved upon and continued, can often seize business opportunities and dominate the market. There are many studies on the levels of the needs of consumers. Nevertheless, what's most important is to analyze and discover the needs of consumers.

The concept of affair-elements in Extenics can be utilized to provide a formalized presentation of needs, to study the extensible property of needs, and to provide formalized analysis methods for needs.

Because the needs of consumers correspond to the functions of products, formalized new product ideas can be created from the needs to help product developers to conceive new products.

8.2.1 The Basic Idea of the First Creation Method

The first creation method is used to generate a creative idea for a brand new product according to the functions consumers need. The brand new product is a product of new technology, new invention, and new discovery that has never been seen before in the world. People's lifestyles may be changed due to the birth of such a new product, leading to "consumption revolution."

In fact, a product is something with certain functions that meet people's needs. Need is the mother of invention. Products (or services) can be created by needs, and the market and enterprises, or even an industry, can be consequently created.

A brand new product can be created if a new product is conceived from the unsatisfied needs of consumers. It's a great way to avoid competition and create blue oceans.

The primary goal of the first creation method is not to directly create new products, but to solve the contradiction of products in the current market that fail to satisfy certain needs of consumers. That is, it is a method to generate creative ideas of brand new products under the situation that consumers' needs of certain functions cannot be satisfied by existing objects in the objective world.

For instance, people's life rhythms have become faster and faster, while the majority of people travel less frequently and go shorter distances since the beginning of the twenty-first century. People prefer to rest at home rather than join the domestic travel wave, even if they have time and financial means during national holidays. In this case, consumers urgently need products that can meet their needs like relaxing, being close to nature, reducing consumption, etc., under the uncrowded condition.

The dream can be fulfilled with "virtual reality glasses" created by sophisticated virtual reality technology. The glasses are a combination of technologies in many aspects, and they allow people to feel like they are participating in the scene by way of vision, hearing, touch, and other aspects. Virtual reality glasses can contribute to solving problems of people being too busy to go on a tour due to a busy lifestyle or people being unwilling to travel due to the crowded tourist destinations. It can be predicted that the technology will change people's way of life in the near future. Recently, virtual reality technology has been widely applied in the medical industry, entertainment, military aerospace, interior design, real estate development, industrial simulation, and other fields.

8.2.2 Key Steps of the First Creation Method

1. **Analyze the needs of consumers and the reason why the needs are unmet, and establish an extension model of the contradictory problem to determine product functions that need to be created.**

   After determining the reason why there is no product that can satisfy the needs, an extension model of the contradictory problem should be established, which should correspond to the functions of product O that need to be created. Affair-elements are adopted to represent these functions to form a set $\{A_f\}$ of functional affair-elements.

2. **Determine the set of property characteristic-elements and the set of real characteristic-elements of the product O that needs to be created.**

   Functions of many products are determined by the properties of the products. For example, a sheet of paper can function as a wrapper because the paper has the property characteristic of "flexibility" and corresponding values. In turn, the property characteristic is determined because the paper has the characteristic of "texture" and corresponding values. The functions of some products are directly determined by the real characteristics of the products. For example, a pen has a function of "pressing the paper" because the pen has the real characteristic of "quality" and its corresponding values.

Given this, it is necessary to determine the set $\{(c_g, v_g)\}$ of property characteristic-elements and the set $\{(c_r, v_r)\}$ of real characteristic-elements after determining the set of functional affair-elements in the first step.

3. **Determine the hard and soft part of the product O to be created.**

   According to the set $\{(c_r, v_r)\}$ of real characteristic-elements, the set $\{(O_r, c_r, v_r)\}$ of real matter-elements is constructed, representing the matter-elements that can be used as components, materials, and raw materials, i.e. the hard part hr$O$ of the product. Then relations among the components, i.e. the soft part sf$O$ of the product, are designed.

4. **Determine the latent and negative parts of the product O that needs to be created.**

   According to the requirements of the product O's latent function, and considering product O's part whose values about certain characteristics are negative, determine the latent part lt$O$ and the negative part ng$_c O$ of the product O.

5. **Extend and transform these conjugate parts with the extensible analysis and extension transformations to obtain new creative ideas for judging whether they can solve the contradictory problem of concern.**

   Extensible analysis and extension transformations can be utilized to obtain new product ideas that meet the needs of consumers so that O has the latent function required. Meanwhile, the negative effects brought by the product can be reduced. Then judge whether the original contradictory problem can be resolved with these ideas. If they are not resolved, the extensible analysis and extension transformations should be repeatedly applied until creative ideas that can resolve the contradictory problem are generated.

6. **Obtain a set of product creative ideas.**

   Thus, a number of creative ideas of the product O are created, i.e. a set $\{B\}$ of the product creative ideas can be obtained from the above steps.

7. **Evaluate.**

   Comprehensive evaluations are conducted on the creative ideas in $\{B\}$ to determine which idea is superior. The above steps are expressed in a flow chart in Figure 8.2.1.

Case Analysis:

Example 8.2.1

A person who will go on a business trip or a journey would like to bring only an carry-on suitcase with several pairs of shoes for adapting to different occasions. Use the first creation method to generate creative ideas of a new product to satisfy the need.

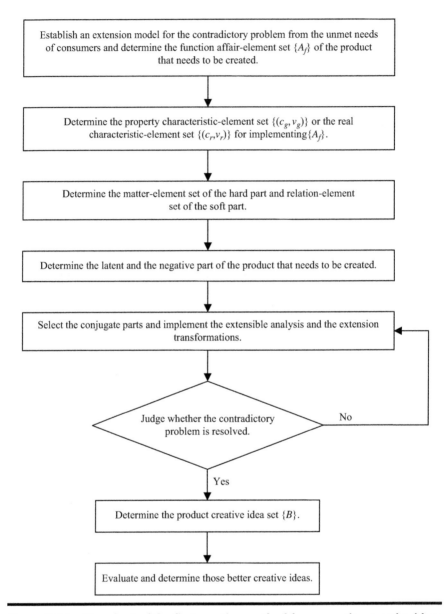

Figure 8.2.1   Flowchart of the first creation method for generating creative ideas of new products.

1. Analyze the needs of the consumer and the reasons that the needs are unmet, and establish an extension model of the contradictory problem to determine the function of the product O that needs to be created.

   Suppose that in this problem, the needs of the consumer are

$$
A(t_i) = \begin{bmatrix} \text{Wear,} & \text{dominating object,} & \{D_i(t_i)\} \\ & \text{acting object,} & \text{travellers} \\ & \text{occasions,} & t_i \end{bmatrix} \wedge \begin{bmatrix} \text{Bring,} & \text{dominating object,} & \{D_i(t_i)\} \\ & \text{tools,} & M_D \end{bmatrix},
$$

$i = 1, 2, 3, 4, 5$, where

$$
M_D = \begin{bmatrix} \text{Suitcase } D, & \text{capacity,} & v \\ & \text{amount,} & \text{one} \end{bmatrix},
$$

$t_1 = $ meeting, $D_1(t_1) = $ leather shoes; $t_2 = $ running, $D_2(t_2) = $ running shoes; $t_3 = $ leisure, $D_3(t_3) = $ leisure shoes; $t_4 = $ climbing, $D_4(t_4) = $ climbing boots; $t_5 = $ playing tennis, and $D_5(t_5) = $ tennis shoes.

The volume of each pair of shoes is, for our purposes, constant, and the capacity of the suitcase is constant if only one carry-on suitcase is allowed. Thus, the problem represents a contradiction constituted by the volume of the suitcase and the volume of all of the shoes carried. That is, we have

$$
P = \left(\{D_i(t_i)\}, \quad \text{volume,} \quad \sum_{i=1}^{5} v_i \right) * \left(\text{Suitcase } D, \quad \text{capacity,} \quad v \right), \ i = 1, 2, 3, 4, 5.
$$

The function of compatibility degree is $k(x) = v - v_0 - x$, where v_0 is the volume of all items other than shoes to be taken by the traveller. For the existing shoes, the sum of the volumes of the five pairs of shoes is $x_0 = \sum_{i=1}^{5} v_i$. Thus,

$$
k(x_0) = v - v_0 - \sum_{i=1}^{5} v_i < 0.
$$

Because the volume v of the suitcase is constant, one approach is to reduce the volume of each pair of shoes. However, all shoes $D_i(t_i)$ are made of soles $D_{1i}(t_i)$, uppers $D_{2i}(t_i)$, and other accessories $D_{3i}(t_i)$. As the size of various kinds of shoes is the same for a person, the size of the soles is the same, and the sole is a part that cannot be contracted and thus takes up the greatest proportion of space in the suitcase.

The functional affair-element that corresponds to $\{D_i(t_i)\}$ is

$$
A_f(t_i) = \begin{bmatrix} \text{Protect,} & \text{dominating object,} & \text{foot}(t_i) \\ & \text{acting object,} & \text{traveller} \\ & \text{occasions,} & t_i \\ & \text{tools,} & \{D_i(t_i)\} \end{bmatrix}
$$

$$
\wedge \begin{bmatrix} \text{Place,} & \text{dominating object,} & \{D_i(t_i)\} \\ & \text{position,} & M_D \end{bmatrix}.
$$

$$
i = 1, \ 2, \ldots, 5.
$$

2. Determine the set of property characteristic elements and the set of real characteristic elements of the product O.

　To achieve the above functions, $\{D_i(t_i)\}$ must have the following set of property characteristic elements:

$$
\{(c_g, v_g)\} = \begin{cases} (\text{foldability, height}), (\text{softness, height}), (\text{comfortable feature, height}), \\ (\text{adaptability, height}) \end{cases}.
$$

To achieve these property characteristic elements, $\{D_i(t_i)\}$ must have the following set of real characteristics elements:

$$
\{(c_r, v_r)\} = \{(\text{volume,} \ v_i), (\text{size,} \ u_{1i}), (\text{weight,} \ u_{2i}), (\text{material,} \ u_{3i}), (\text{color,} \ u_{4i})\},
$$

$$
i = 1,2,3,4,5.
$$

3. Determine the hard and soft parts of the product O.

Because the product O is designed to protect feet in a variety of different situations, according to general knowledge, the hard parts of the product must be equipped with "sole $D_{1i}(t_i)$", "upper $D_{2i}(t_i)$" and "accessories $D_{3i}(t_i)$". In addition, the sizes of the products needed for different occasions should be same for our purposes, shown symbolically as

$$
(\text{size,} \ u_{1i}) = (\text{size,} \ u_1).
$$

According to these real characteristic elements, the matter-elements of the hard parts can be constructed as follows:

$$
M_{1hr}(t_i) = \begin{bmatrix} D_{1i}(t_i), & \text{volume,} & v_{1i} \\ & \text{size,} & u_1 \\ & \text{weight,} & u_{21i} \\ & \text{material,} & u_{31i} \\ & \text{color,} & u_{41i} \end{bmatrix}, \quad M_{2hr}(t_i) = \begin{bmatrix} D_{2i}(t_i), & \text{volume,} & v_{2i} \\ & \text{size,} & u_1 \\ & \text{weight,} & u_{22i} \\ & \text{material,} & u_{32i} \\ & \text{color,} & u_{42i} \end{bmatrix},
$$

$$
M_{3hr}(t_i) = \begin{bmatrix} D_{3i}(t_i), & \text{volume,} & v_{3i} \\ & \text{size,} & u_1 \\ & \text{weight,} & u_{23i} \\ & \text{material,} & u_{33i} \\ & \text{color,} & u_{43i} \end{bmatrix}
$$

In addition, the matter-elements of the hard parts must satisfy $\sum_{i=1}^{5}\sum_{j=1}^{3} v_{ji} = \sum_{i=1}^{5} v_i < v - v_0$.

For simplicity, only the soft parts consisting of the sole and upper are analyzed in this case. Relation-elements of the soft parts consisting of soles $D_{1i}(t_i)$ and uppers $D_{2i}(t_i)$ of the existing shoes are

$$
R_{sf}(t_i) = \begin{bmatrix} \text{Connection relation,} & \text{antecedent,} & D_{2i}(t_i) \\ & \text{consequent,} & D_{1i}(t_i) \\ & \text{degree,} & \text{tightness} \\ & \text{means,} & \text{glue} \end{bmatrix}.
$$

4. Determine the latent and negative parts of the product to be created.

According to the consumer's needs, the products composed of the above hard parts and soft parts must meet the volume requirement. Thus, it is unnecessary to consider the latent parts of the product. With respect to the product O's volume, the soles and uppers belong to the negative parts (for shoes, the volume of accessories can be ignored) and cannot directly accept any contraction transformation.

5. Appropriately perform the extensible analysis and apply an extension transformation to determine whether the transformed objects translate the originally incompatible problem into a compatible one.

According to the methods of conjugate analysis and conjugate transformation, a divergence analysis is first conducted on the soft parts as follows:

$$
R_{sf}(t_i) - \left\{
\begin{array}{l}
R_{sf1}(t_i) = \begin{bmatrix}
\text{Connection relation,} & \text{antecedent,} & D_{2i}(t_i) \\
 & \text{consequent,} & D_{1i}(t_i) \\
 & \text{degree,} & \text{tightness} \\
 & \text{means,} & \text{traditional thread binding}
\end{bmatrix} \\[2em]
R_{sf2}(t_i) = \begin{bmatrix}
\text{Connection relation,} & \text{antecedent,} & D_{2i}(t_i) \\
 & \text{consequent,} & D_{1i}(t_i) \\
 & \text{degree,} & \text{tightness} \\
 & \text{means,} & \text{bind}
\end{bmatrix} \\[2em]
R_{sf3}(t_i) = \begin{bmatrix}
\text{Connection relation,} & \text{antecedent,} & D_{2i}(t_i) \\
 & \text{consequent,} & D_{1i}(t_i) \\
 & \text{degree,} & \text{tightness} \\
 & \text{means,} & \text{zipper}
\end{bmatrix}
\end{array}
\right.
$$

If the following active transformation φ is implemented,

$$
\varphi R_{sf}(t_i) = \begin{bmatrix}
\text{Connection relation,} & \text{antecedent,} & D_{2i}(t_i) \\
 & \text{consequent,} & D_{1i}(t_i) \\
 & \text{degree,} & \text{tightness} \\
 & \text{means,} & \text{zipper}
\end{bmatrix} = R_{sf3}(t_i),
$$

then the following active transformation φ' is also implemented:

$$
\varphi' M_{1hr}(t_i) = \begin{bmatrix}
D_1, & \text{volume,} & v_1 \\
 & \text{size,} & u_1 \\
 & \text{weight,} & u_{21} \\
 & \text{material,} & u_{31} \\
 & \text{color,} & u_{41}
\end{bmatrix} = M_{1hr}.
$$

That is, all soles are changed to a fixed sole that is not changed along with t_i. The following conductive transformation of the soft parts can be obtained:

$$T_{i\varphi'\varphi}R_{sf3}(t_i) = \begin{bmatrix} \text{Connection relation,} & \text{antecedent,} & D_{2i}(t_i) \\ & \text{consequent,} & D_1 \\ & \text{degree,} & \text{tightness} \\ & \text{means,} & \text{zipper} \end{bmatrix} = R_{sf3}{}'(t_i), \ i = 1, 2, \ldots, 5.$$

Apparently, $x' = v_1 + \sum_{i=1}^{5} v_{2i} + \sum_{i=1}^{5} v_{3i} < v - v_0$. The compactibility degree after the transformation is $k(x') = v - v_0 - x' > 0$, i.e., the incompatible problem is transformed into a compatible one.

6. Determine the set $\{B\}$ of product creative ideas.

   According to these transformations, multiple product creative ideas can be determined:

$$\{B\} = M_{1hr} \wedge \{M_{2hr}(t_i)\} \wedge \{M_{3hr}(t_i)\} \wedge R'_{sf3}(t_i).$$

The meaning of these ideas is that a fixed sole is adopted with five kinds of different uppers and accessories. The sole is connected with different uppers by means of a zipper connection, so as to constitute various product ideas meeting the requirement of volume.

Illustration: There are various relations between the upper and accessories as well as between accessories and the sole in the actual design. These relations are not analyzed because they are irrelevant to the conversion of the contradictory problem in this case study.

7. Evaluation and superiority selection.

   The above product ideas are evaluated with (comfortable feature, high), (adaptability, high) as measuring indicators; the details are omitted here. The final choice is the idea of a casual slope sole with five kinds of different uppers and accessories where zippers are used to connect the sole and the upper.

8.3 The Second Creation Method: Generating Creative Ideas of Products Based on Existing Products

QUESTIONS AND THINKING:

- How can we conceive new products or series of products from existing products?
- What characteristics, functions, and structures do the existing products have?
- Which point of view do you tend to conceive new products from?
- How do you get your ideas?

To compete for market share and obtain a larger share of the "cake," many companies have presented their most creative and outstanding ideas by focusing on several characteristics of their products, leading to high spending on these characteristics. To create new products, the extensible analysis method can be utilized to extend other values of these characteristics or other characteristics. By doing so, we can not only create distinctive products, but we can also create new products through four basic transformations or their operations based on characteristics of these products.

For instance, Eastman Kodak dominated the landscape in the film era. Other camera manufacturers could get only a small portion of the market from Kodak, no matter how they innovated their products. However, with the advent of digital media, other companies have continuously innovated their digital technologies. Fujitsu, Canon, and other brands of digital cameras have appeared in the market. Digital cameras were able to capture a huge market share very quickly due to features of usability and efficiency. Smart phones have also occupied part of the card-type digital camera market.

The second creation method, which is commonly utilized to generate a series of products, is to extend and transform certain elements of one or several existing products to generate new product ideas.

The main steps of the second creation method are presented here.

1. Decompose the original product: The original product O is decomposed through conjugate analysis to list the material and nonmaterial parts, soft and hard parts, latent and apparent parts, as well as negative and positive parts. Their intermediate parts should be also listed if possible. One or multiple aspects of materiality, systemicness, dynamism, or antagonism should be determined for innovation.

2. List the main characteristics of the original product O and its soft and hard conjugate parts, and obtain the values corresponding to the above characteristics according to the actual problems. Also list the main functions of the product O.

3. List the matter-elements (including the whole product, its components, parts, etc.) of the hard part of the original product O at different levels and relation-elements of various kinds of the soft part (including structures of the product, various connected relations, etc.). Then list the functional affair-elements of the hard and soft parts of the original product O at different levels.

4. Conduct an extensible analysis: Use the extensible analysis method to extend the above basic-elements to obtain multiple paths of innovation.

5. Conduct an extension transformation: Transform basic-elements in step 3 into various basic-elements extended in step (4) using various extension

transformation methods. Conduct operations of extension transformations to obtain a variety of new product ideas.

6. Evaluation and superiority selection: Evaluate new product ideas formed from the extension transformations by using the superiority evaluation method. The ideas with higher superiority can be regarded as new product ideas to be selected.

The above steps can be expressed in the flow chart in Figure 8.3.1.

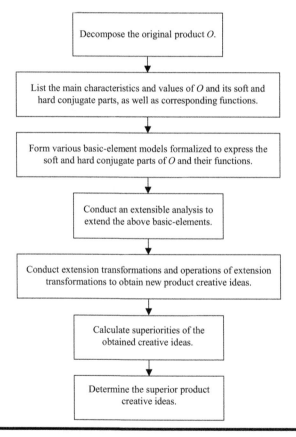

Figure 8.3.1 Flowchart of the second creation method for generating creative ideas of new products.

Case Analysis:

Example 8.3.1

Utilize the second creation method to obtain new product ideas from a certain existing square mahogany table O.

1. The existing square mahogany table O is decomposed from the perspective of the soft and hard conjugate analysis. Table O can be decomposed into table top O_1 and table legs O_2. Some other important characteristics of the table and corresponding values are listed, forming product matter-elements, structural relation-elements, and functional affair-elements as follows:

$$
M = \begin{bmatrix} O, & \text{amount of legs,} & 4 \\ & \text{material,} & \text{bloodwood} \\ & \text{price,} & 10\ 000\ \text{yuan} \\ & \text{divisibility,} & \text{none} \end{bmatrix}, \quad
M_1 = \begin{bmatrix} O_1, & \text{shape,} & \text{square} \\ & \text{material,} & \text{bloodwood} \\ & \text{price,} & 8\ 000\ \text{yuan} \\ & \text{area,} & 2.25\ \text{m}^2 \\ & \text{divisibility,} & \text{none} \end{bmatrix},
$$

$$
M_2 = \begin{bmatrix} O_2, & \text{shape,} & \text{cuboid} \\ & \text{material,} & \text{bloodwood} \\ & \text{price,} & 2\ 000\ \text{yuan} \\ & \text{height,} & 1.2\ \text{m} \\ & \text{divisibility,} & \text{none} \end{bmatrix},
$$

$$
R = \begin{bmatrix} \text{Connection relation,} & \text{antecedent,} & O_1 \\ & \text{consequent,} & O_2 \\ & \text{means,} & \text{screw} \\ & \text{degree,} & \text{tightness} \end{bmatrix},
$$

$$
A_1 = \begin{bmatrix} \text{Place,} & \text{dominating object,} & \text{cutlery} \\ & \text{acting object,} & \text{old people} \\ & \text{tools,} & O \\ & \text{occasions,} & \text{dining hall} \end{bmatrix}
$$

$$
A_2 = \begin{bmatrix} \text{Promote,} & \text{dominating object,} & \text{taste} \\ & \text{acting object,} & \text{old people} \\ & \text{tools,} & O \\ & \text{occasions,} & \text{dining hall} \end{bmatrix}
$$

2. Conduct an extensible analysis on various basic elements of the original product O (for simplicity, extensions of the functional affair-elements are not conducted). By taking the divergence analysis as an example, we can obtain the following trees.

$$
M = \begin{bmatrix}
O, & \text{number of legs,} & 4 \\
& \text{material,} & \text{bloodwood} \\
& \text{price,} & \text{10 000 yuan} \\
& \text{divisibility,} & \text{none}
\end{bmatrix}
$$

$$
M^1 = \begin{bmatrix}
O^1, & \text{number of legs,} & 1 \\
& \text{material,} & \text{bloodwood} \\
& \text{price,} & \text{8 000 yuan} \\
& \text{divisibility,} & \text{none}
\end{bmatrix}
\dashv M^{11} = \begin{bmatrix}
O^1, & \text{color,} & \text{bloodwood} \\
& \text{weight,} & a_{11} \\
& \text{foldability,} & \text{none}
\end{bmatrix}
$$

$$
M^2 = \begin{bmatrix}
O^2, & \text{number of legs,} & 3 \\
& \text{material,} & \text{oak} \\
& \text{price,} & \text{9 000 yuan} \\
& \text{divisibility,} & \text{exist}
\end{bmatrix}
\dashv M^{21} = \begin{bmatrix}
O^2, & \text{color,} & \text{oak} \\
& \text{weight,} & a_{21} \\
& \text{foldability,} & \text{none}
\end{bmatrix}
$$

$$
M^3 = \begin{bmatrix}
O^3, & \text{number of legs,} & 3 \\
& \text{material,} & \text{black walnut} \\
& \text{price,} & \text{10 000 yuan} \\
& \text{divisibility,} & \text{none}
\end{bmatrix}
\dashv M^{31} = \begin{bmatrix}
O^3, & \text{color,} & \text{dark walnut} \\
& \text{weight,} & a_{31} \\
& \text{foldability,} & \text{exist}
\end{bmatrix}
$$

$$
M^4 = \begin{bmatrix}
O^4, & \text{number of legs,} & 4 \\
& \text{material,} & \text{bloodwood} \\
& \text{price,} & \text{9 000 yuan} \\
& \text{divisibility,} & \text{none}
\end{bmatrix}
$$

$$M_1 = \begin{bmatrix} O_1, & \text{shape,} & \text{square} \\ & \text{material,} & \text{bloodwood} \\ & \text{price,} & \text{8 000 yuan} \\ & \text{area,} & 2.25 \text{ m}^2 \\ & \text{divisibility,} & \text{none} \end{bmatrix}$$

$$\dashv \left\{ \begin{array}{l} M_{11} = \begin{bmatrix} O_{11}, & \text{shape,} & \text{rectangle} \\ & \text{material,} & \text{bloodwood} \\ & \text{price,} & \text{8 000 yuan} \\ & \text{area,} & 2.0 \text{ m}^2 \\ & \text{divisibility,} & \text{none} \end{bmatrix} \dashv M_{111} = \begin{bmatrix} O_{11}, & \text{color,} & \text{rose wood} \\ & \text{foldability,} & \text{none} \\ & \text{thickness,} & b_{11} \end{bmatrix} \\[3em]

M_{12} = \begin{bmatrix} O_{12}, & \text{shape,} & \text{square} \\ & \text{material,} & \text{oak} \\ & \text{price,} & \text{8 000 yuan} \\ & \text{area,} & 2.25 \text{ m}^2 \\ & \text{divisibility,} & \text{exist} \end{bmatrix} \dashv M_{121} = \begin{bmatrix} O_{12}, & \text{color,} & \text{oak} \\ & \text{foldability,} & \text{exist} \\ & \text{thickness,} & b_{12} \end{bmatrix} \\[3em]

M_{13} = \begin{bmatrix} O_{13}, & \text{shape,} & \text{petal-shaped} \\ & \text{material,} & \text{bloodwood} \\ & \text{price,} & \text{10 000 yuan} \\ & \text{area,} & 1.8 \text{ m}^2 \\ & \text{divisibility,} & \text{exist} \end{bmatrix} \dashv M_{131} = \begin{bmatrix} O_{13}, & \text{color,} & \text{rose wood} \\ & \text{foldability,} & \text{exist} \\ & \text{thickness,} & b_{13} \end{bmatrix} \\[3em]

M_{14} = \begin{bmatrix} O_{14}, & \text{shape,} & \text{hexagon} \\ & \text{material,} & \text{black walnut} \\ & \text{price,} & \text{8 000 yuan} \\ & \text{area,} & 1.8 \text{ m}^2 \\ & \text{divisibility,} & \text{none} \end{bmatrix} \dashv M_{141} = \begin{bmatrix} O_{14}, & \text{color,} & \text{dark walnut} \\ & \text{foldability,} & \text{none} \\ & \text{thickness,} & b_{14} \end{bmatrix} \end{array} \right.$$

$$
M_2 = \begin{bmatrix} O_2, & \text{shape}, & \text{cuboid} \\ & \text{material}, & \text{bloodwood} \\ & \text{price}, & 2\,000 \text{ yuan} \\ & \text{height}, & 1.2 \text{ m} \\ & \text{divisibility}, & \text{none} \end{bmatrix} \dashv \begin{cases} M_{21} = \begin{bmatrix} O_{21}, & \text{shape}, & \text{cylinder} \\ & \text{material}, & \text{stainless steel} \\ & \text{price}, & 3\,000 \text{ yuan} \\ & \text{height}, & 1.2 \text{ m} \\ & \text{divisibility}, & \text{none} \end{bmatrix} \\[1em] M_{22} = \begin{bmatrix} O_{22}, & \text{shape}, & \text{cuboid} \\ & \text{material}, & \text{aluminum alloy} \\ & \text{price}, & 12\,000 \text{ yuan} \\ & \text{height}, & 1.2 \text{ m} \\ & \text{divisibility}, & \text{exist} \end{bmatrix} \\[1em] M_{23} = \begin{bmatrix} O_{23}, & \text{shape}, & \text{cuboid} \\ & \text{material}, & \text{black walnut} \\ & \text{price}, & 2\,000 \text{ yuan} \\ & \text{height}, & 1.0 \text{ m} \\ & \text{divisibility}, & \text{exist} \end{bmatrix} \end{cases}
$$

$$
R = \begin{bmatrix} \text{Connection relation}, & \text{antecedent}, & O_1 \\ & \text{consequent}, & O_2 \\ & \text{means}, & \text{screw} \\ & \text{degree}, & \text{strong} \end{bmatrix}
$$

$$
\dashv \begin{cases} R_1^{ij} = \begin{bmatrix} \text{Connection relation}, & \text{antecedent}, & O_{1i} \\ & \text{consequent}, & O_{2j} \\ & \text{means}, & \text{an organic whole} \\ & \text{degree}, & \text{tightness} \end{bmatrix} \\[1em] R_2^{ij} = \begin{bmatrix} \text{Connection relation}, & \text{antecedent}, & O_{1i} \\ & \text{consequent}, & O_{2j} \\ & \text{means}, & \text{embedded type} \\ & \text{degree}, & \text{tightness} \end{bmatrix} \\[1em] R_3^{ij} = \begin{bmatrix} \text{Connection relation}, & \text{antecedent}, & O_{1i} \\ & \text{consequent}, & O_{2j} \\ & \text{means}, & \text{screw type} \\ & \text{degree}, & \text{tightness} \end{bmatrix} \end{cases}
$$

$$
(i = 1, 2, 3, 4;\ j = 1, 2, 3)
$$

Extension transformations and operations of extension transformations can be conducted on each matter-element and each relation-element to obtain a variety of new product ideas. Due to the limitation of space, the following extension transformations and their AND operations of transformations are taken as an example.

- Conduct extension transformations: $T^3M = M^3 \oplus M^{31}$, $T_{14}M_1 = M_{14} \oplus M_{141}$, $T_{23}M_2 = M_{23}$, and $T_{R1}R = R_1^{43}$. The composite transformation of these transformations is

$$T_1 = T_{R1}\left(T^3 \wedge T_{14} \wedge T_{23}\right).$$

- Conduct extension transformations: $T_{13}M_1 = M_{13}$, $T_{21}M_2 = M_{21}$, and $T_{R3}R = R_3^{31}$, . The AND transformation $T_{13} \wedge T_{21}$ will lead to the following conductive transformation.

$$
{}_{13 \wedge 21}T^5M = M^5 = \begin{bmatrix} O^5, & \text{number of legs,} & 4 \\ & \text{material,} & \text{bloodwood} \\ & \text{price,} & \text{13 000 yuan} \\ & \text{divisibility,} & \text{exist} \end{bmatrix},.
$$

So, the composition transformation of all these transformations is

$$T_2 = {}_{13 \wedge 21}T^5T_{R3}\left(T_{13} \wedge T_{21}\right).$$

The extension models of the new product obtained from the above transformations are

$$M_*^3 = \left(M^3 \oplus M^{31}\right) \wedge \left(M_{14} \oplus M_{141}\right) \wedge M_{23} \wedge R_1^{43} \text{ and }$$

$$M_*^5 = M^5 \wedge M_{13} \wedge M_{21} \wedge R_3^{31}$$

These two idea models can be expressed in words as follows:

- **Creative idea 1:** A black walnut table with three rectangular legs and a one-piece indivisible hexagonal tabletop, priced at 10,000 yuan.
- **Creative idea 2:** A table with four cylindrical stainless steel legs and a split petal-shaped mahogany tabletop, priced at 13,000 yuan. For example, the tabletop can be split into multiple pieces that can be combined into a whole "petal-shaped" table, greatly increasing the flexibility of the table. The table is especially suitable for occasions of individual use and frequently moving.

Similarly, additional ideas can be generated. The superiority evaluation method should then be utilized to evaluate and obtain the superior ideas according to the technical feasibility of the particular enterprise, market demands, costs, and other situations, which are all omitted here.

8.4 The Third Creation Method: Generating Creative Ideas of Products Based on Products' Defects

QUESTIONS AND THINKING:

- Do you often complain about the products you use?
- How do you think about improving these products? From what perspective do you usually consider making improvements?
- How do you get your ideas?

For a product, comments from customers who are unwilling to buy the product or who are dissatisfied with their purchase have been investigated. Upon analysis, shortcomings in the product should be found and improved. Thus, a new product can be created. On the one hand, the original customer demands for the new product should be maintained. That is to say, customers of the new product consist of two types: (1) some of them are non-customers of the original product; (2) some of them are customers of the original product. The market perspective of the new product is better than that of the old one.

The third creation method is to conduct an extensible analysis and extension transformations on shortcomings or customers' complaints of the existing products in the marketplace to generate creative new product ideas. Superiority analysis must also be conducted on the creative ideas generated to select a superior creative idea. The main steps of this third creation method are presented as follows.

1. Upon analyzing the existing products in the marketplace, with models of product matter-elements, functional affair-elements, and structural relation-elements, those products can be formally expressed as

$$
M = \begin{bmatrix} O_M, & c_{M01}, & v_{M01} \\ & c_{M02}, & v_{M02} \\ & \vdots & \vdots \\ & c_{M0m_1}, & v_{M0m_1} \end{bmatrix}, A = \begin{bmatrix} O_A, & c_{A01}, & v_{A01} \\ & c_{A02}, & v_{A02} \\ & \vdots & \vdots \\ & c_{A0n_2}, & v_{A0n_2} \end{bmatrix}, R = \begin{bmatrix} O_R, & c_{R01}, & v_{R01} \\ & c_{R02}, & v_{R02} \\ & \vdots & \vdots \\ & c_{R0n_3}, & v_{R0n_3} \end{bmatrix}
$$

where O_M represents an existing product to be analyzed, c_{M0i} a certain characteristic of the product, v_{M0i} the value of the product with respect to the characteristic, O_A the action in the functions of the product to be analyzed, c_{A0j} a certain characteristic of the action, v_{A0j} the value of the action with respect to the characteristic c_{A0j}, O_R the relational word in the structure of the product to be analyzed, c_{R0l} a certain characteristic of the relational word, and v_{R0l} the value of the relational word with respect to the characteristic c_{R0l}.

If the analysis on the shortcomings of the product's entity is emphasized, then the product matter-elements are analyzed. If the analysis on the functional shortcomings is emphasized, then functional elements are analyzed. If the analysis on the structural shortcomings is emphasized, then structural elements are analyzed. These elements can be analyzed either separately or simultaneously.

2. According to these extension models, the shortcomings listing method is adopted to find shortcomings of the product. Characteristics of the shortcomings of the product, characteristics of functional shortcomings, and characteristics of structural shortcomings are listed separately to rewrite models, allowing the characteristics in the model to correspond to the shortcoming characteristics of different degrees from top to bottom. In this way, requirements of the shortcomings can be analyzed. The rewritten extension model is presented as

$$M_0 = \begin{bmatrix} M_{01} \\ \vdots \\ M_{0m_1} \\ \vdots \\ M_{0m_1} \end{bmatrix}, \quad A_0 = \begin{bmatrix} A_{01} \\ \vdots \\ A_{0m_2} \\ \vdots \\ A_{0n_2} \end{bmatrix}, \quad \text{and} \quad R_0 = \begin{bmatrix} R_{01} \\ \vdots \\ R_{0m_3} \\ \vdots \\ R_{0n_3} \end{bmatrix},$$

where $M_{01}, M_{02}, \ldots, M_{0m_1}$ are the shortcomings' sub-matter-elements with respect to the shortcomings' characteristics of the product entity, $A_{01}, A_{02}, \ldots, A_{0m_2}$ are the shortcomings' sub-affair-elements with respect to the shortcomings' characteristics of the product's functions, and $R_{01}, R_{02}, \ldots, R_{0m_3}$ are the shortcomings' sub-relation-elements with respect to the shortcomings' characteristics of the product structures.

In general, when the analysis is conducted manually, modeling is conducted directly starting from step 2 without the need to list all of the divided non-shortcoming basic-elements.

3. An extensible analysis is conducted on the obtained shortcomings' sub-basic-elements using divergent analysis, correlation analysis, implication analysis, or expandable analysis.

4. According to these analysis results, the extension transformation method is utilized to carry out transformations or operations of transformations on the

shortcomings' sub-basic-elements. Results of conductive transformations are also considered if there is any relevant basic-element so that several new product ideas can be formed comprehensively.

5. According to market surveys and historical data, measuring indicators are determined. A superiority evaluation is conducted on the new product ideas generated above from the perspective of social indicators, economic indicators, technical indicators, and so on, so as to select an idea with high superiority.

Note: Because functions and structures are interrelated to many characteristics of the product, correlation analyses and implication analyses should be applied in the extension process to ensure that the conductive transformations occurring in the process of implementing the active extension can be fully considered, preventing situations from getting worse.

A simplified process of the third creation method is shown in Figure 8.4.1.

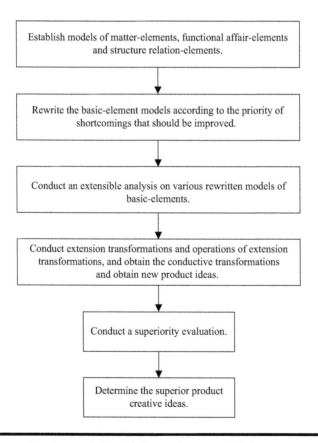

Figure 8.4.1 Flowchart of the third creation method for generating creative ideas of new products.

Case Analysis:

Example 8.4.1

An existing laser pointer with a red light and a black rectangular housing is priced at 210 yuan. The power source is a AAA battery, and there is no power switch. The laser indicator has three available indicating shapes: dot, straight line segment, and circle. The laser pointer can be used for page up/down, blank screen, and returning directly to the editing state. A USB is positioned at the bottom of the reverse side of the laser pointer. According to user surveys, recognized shortcomings of the product include: ugly shape, monotonous color, no pattern, high power consumption, and high price. Generate new product ideas from the shortcomings of this product.

The third creation method is used below to demonstrate how to obtain new product ideas based on a product's shortcomings. For simplicity, shortcoming matter-elements of the product, shortcoming functional affair-elements, and shortcoming structural relation-elements are listed. For a better understanding, a data table with multi-dimensional basic-elements is adopted as the expression in this example.

1. **Modeling:** Shortcomings of the laser pointer are analyzed and presented in Table 8.4.1 with multi-dimensional matter-element data table.

   There is no functional shortcoming in the laser pointer. However, the high cost of the laser pointer is due to the variety of functions of the laser pointer. Therefore, the main functions of the product should be analyzed to improve the original product. These main functions are presented in Table 8.4.2 with multi-dimensional affair-element data.

   Structural shortcomings of the laser pointer are analyzed and expressed in Table 8.4.3 with multi-dimensional relation-elements data.

Table 8.4.1 Shortcoming Matter-elements of the Laser Pointer

Object	Characteristics	Value
Laser pointer	shape	cuboid
	exterior color	black
	price	210 yuan
	pattern	no
	color of the radial	red
	power source type	AAA battery
	power consumption	high

Table 8.4.2   Functional Affair-elements of the Laser Pointer

Object	Characteristics	Value
Send out	dominating object	red light
	tool	diode
Indicate	dominating object	screen
	tool	indicate key
	status	{dot, line, circle}
Convert	dominating object	instruction status
	tool	runner
Turn over	dominating object	page
	tool	page up/down
	means	{forward, backward}
Return	dominating object	{editing state, broadcasting state}
	tool	return key
	means	press one time
Cover	dominating object	screen
	tool	cover key

Table 8.4.3   Shortcoming Structure Relation-elements of the Laser Pointer

Object	Characteristics	Value
Cutting relation	antecedent	USB card
	consequent	USB port
	degree	tightness
	means	directly embedded

2. **Extensible analysis:** Divergent analysis, correlative analysis, implication analysis, or expandable analysis can be utilized to extend the sub-basic-elements of the listed shortcomings. To understand this, the example is presented in a table of data corresponding to the basic-elements.
   a. An extensible analysis is conducted on the matter-elements of the product's shortcomings. First, divergent analysis as shown in Table 8.4.4 is conducted on the product's matter-elements.

Table 8.4.4 Divergent Analysis of Product Matter-elements

Object	Characteristics	Value	Divergent Results of Value
Laser pointer	shape	cuboid	cube, elliptical cylinder, sphere, cylinder, ball-crown body, animal type body...
	exterior color	black	red, white, blue, full colors,...
	price	210 yuan	[30,400] yuan
	pattern	no	blue -and-white, traditional Chinese paintings, animals, geometric patterns, animals...
	color of the radial	red	green, blue, yellow, purple...
	power source type	AAA battery	AA battery, button battery, lithium battery, rechargeable battery...
	power consumption	high	medium, low...
Divergent Results of Object	Divergent Results of Characteristic	Value	Divergent Results of Value
Power switch	exterior material	plastic	stainless steel, copper...
	shape	elliptical button	cylindrical button
	position	reverse side	the side, front
Mouse	cost	20 yuan	[10,20] yuan
Thermometer	weight	30 g	[20,50] g
Brush	length	12 cm	[10,20] cm

According to the correlation analysis, it can be observed that the high price of the laser pointer is caused by the high cost. According to the expandable analysis, the cost of the laser pointer can be decomposed into the costs of each component, which will demonstrate that the high cost of the laser pointer is caused by the high cost of each functional component. According to the implication analysis, it can be observed that the high power consumption of the laser pointer is due to the fact that no power switch is set.

b. An extensible analysis is conducted on the relation-elements of structural shortcomings. Based on user surveys, users mainly complained about the inconvenience of the USB port. The USB card was difficult to remove from the USB port, and the USB card was easily damaged through this use. Moreover, according to the above implication analysis, it can be observed that a switch relation between the power source and the switch should be added. Therefore, an extensible analysis should be conducted on the structural relation-elements, as shown below, so as to extend more structural relations.

First, divergent analysis, as shown in Table 8.4.5, is conducted on the shortcomings of the product's structure.

3. **Implement extension transformations to generate new product ideas.** According to the initial basic-element table and extented basic-element table, we can obtain a lot of product ideas through implementing extension transformations. The extension transformations and the results are shown in Tables 8.4.6 to 8.4.9.

According to the transformations in Tables 8.4.6 to 8.4.9, the following three creative ideas can be obtained:

Creative idea S_1: An elliptical cylinder laser pointer in a blue housing with a traditional Chinese painting in red light and AAA battery, conducting a decreasing transformation on the indication state and cover screen. An

Table 8.4.5 Divergent Analysis of Structure Relation-elements

Object	Characteristics	Value	Divergent Results of Values
Cutting relation	antecedent	USB card	switch
	consequent	USB port	power source
	degree	tightness	
	way	directly embedded	push, screw, adsorb…
Divergent Results of Object	*Divergent Results of Characteristic*	*Value*	*Divergent Results of Values*
Switch relation	direction		from left to right, from front to back, from bottom to top…
	position		the middle part of the reverse side, the bottom part of the reverse side, the middle part of the side, the bottom part of the side

Table 8.4.6 Product Matter-element Transformations (1)

Before Transformation			Transformation Type	After Transformation		
Object	Characteristics	Value		Object	Characteristics	Value
Laser pointer	shape	cuboid	Substituting transformations of values	Laser pointer	shape	elliptical cylinder
	exterior color	black			exterior color	white
	price	210 yuan			price	210 yuan
	pattern	no			pattern	traditional Chinese painting
	power source type	AAA battery			power source type	AAA battery
	power consumption	high			power consumption	high

Table 8.4.7 Product Matter-element Transformations (2)

Before Transformation			Transformation Type	After Transformation			
Object	Characteristics	Value		Object	Characteristics	Value	
Laser pointer	shape	cuboid	Substituting transformations of the value and adding transformations of the object	Laser pointer ⊕ Power switch	shape	(elliptical cylinder V ball-crown body V beetle body) ⊕ elliptical button	
	exterior color	black	Substituting transformations of values		exterior color	ancient linen V white V full colors	
	pattern	no			pattern	traditional Chinese painting V blue-and-white V beetle	
	power source type	AAA battery			power source type	button buttery V AA battery	
	price	210 yuan	Conductive transformations		price	180 yuan	
	power consumption	high			power consumption	low	
			Adding transformations of the characteristics-element		position of power switch	the bottom of the reverse side	

Table 8.4.8 Functional Affair-element Transformations

	Before Transformation		Transformation Type	After Transformation		
Object	Characteristics	Value		Object	Characteristics	Value
Send out	dominating object	red light	Substitution transformation of value	Send out	dominating object	blue light
	tool	diode			tool	diode
Indicate	dominating object	screen	Removing transformation of value	Indicate	dominating object	screen
	tool	indication key			tool	indication key
	Status	{dot, line, circle}			status	dot
Convert	dominating object	instruction status	Removing transformation of affair-element			
	tool	runner				
Turn over	dominating object	page	No transformation	Turn over	dominating object	page
	tool	page up/down			tool	page up/down
	means	{forward, backward}			means	{forward, backward}
Return	dominating object	{editing state, broadcasting state}	Removing transformation of value	Return	dominating object	{editing state, broadcasting state}
	tool	return key			tool	return key
	way	press one time and press once again			way	press one time
Cover	dominating object	screen	Removing transformation of affair-element			
	tool	cover key				

Table 8.4.9 Structure Relation-element Transformations

Before Transformation			Transformation Type	After Transformation		
Object	Characteristics	Value		Object	Characteristics	Value
Cutting relation	antecedent	USB card	Adding transformation of the object Adding transformation of value	Cutting relation ⊕ Switch relation	antecedent	USB card ⊕ switch
	consequent	USB slot	Adding transformation of value		consequent	USB slot ⊕ power source
	degree	tightness	No transformation		degree	tightness
	way	directly embedded	Substituting transformation of value		means	push
			Adding transformation of the characteristic element		direction	from top to bottom
			Adding transformation of the characteristic element		position	the bottom of the reverse side

elliptical button-type power switch is added at the bottom of the reverse side of the laser pointer. The pointer is priced at 170 yuan.

Creative idea S_2: A ball crown-shaped laser pointer in a white housing with a blue-and-white pattern in red light and a button battery, conducting a decreasing transformation on the indication state and cover screen. An elliptical button-type power switch is added at the bottom of the reverse side of the laser pointer. The pointer is priced at 180 yuan.

Creative idea S_3: Beetle-shaped laser pointer with a colorful beetle pattern in red light, a AA battery, conducting a decreasing transformation on the indication state and cover screen. The power of the pointer can be switched off by pushing the slot from bottom to top located at the bottom of the reverse side of the laser pointer and can be turned on by fetching out the slot. The pointer is priced at 190 yuan.

By following this method of analysis, more ideas can be obtained, which will not be described in details.

4. **Evaluation and superiority selection of creative ideas.** The ideas obtained above can be evaluated by using the superiority evaluation method to obtain the superior ideas to improve the shortcomings. Specific steps are presented here.

a. Determine measuring indicators: According to market surveys within the customer base of the laser pointer, three measuring indicators (i.e. price, aesthetic degree, and technology innovating degree) should be determined. The set of measuring indicators is regarded as $MI = \{MI_1, MI_2, MI_3\}$, where

$$MI_1 = (\text{price, the price range consumers can accept}) = (\text{price, } [150, 230] \text{ yuan}),$$

$$MI_2 = (\text{aesthetic degree, good or very good}), \quad \text{and}$$

$$MI_3 = (\text{technology innovating degree, above the level of incremental innovation}).$$

b. Determine weight coefficients: According to the results of market surveys and historical data, the priorities of the above three indicators are divided to determine the importance of the different indicators. For example, the weight coefficients of MI_1, MI_2, MI_3 are considered to be $\alpha_1 = 0.3, \alpha_2 = 0.3$, and $\alpha_3 = 0.4$, respectively.

c. Establish dependence functions and calculate the dependence degree: A dependence function is adopted to measure and evaluate the degree of characteristics that satisfy the given requirements.

Dependence functions of costs: According to a survey, the acceptable range of the product's cost for the customer base is 150–230 yuan, expressed as the positive domain of X_1; the satisfied cost range is 150–200 yuan, expressed as standard positive domain X_{01}; and the most satisfied cost is 150 yuan, expressed as x_{01}. Considering the product's quality and consumers' consumption ability, the laser pointer's cost either below 30 yuan or above 300 yuan are rarely accepted by the customer base, and the costs within the ranges of [140,150] and [230,250] can be acceptable with certain promotional means. The range is called the transitional negative domain. The positive transitional domain is combined with the positive domain, expressed as $\hat{X}_1$. Thus, based on the above descriptions, ranges of cost of different degrees can be divided into the following intervals:

$$X_{01} = \left[a_{01}, b_{01}\right] = \left[150, 200\right], X_1 = \left[a_1, b_1\right] = \left[150, 230\right], \hat{X}_1 = \left[c_1, d_1\right] = \left[140, 250\right].$$

It can be seen that the optimal point x_{01} is the left end point of X_{01}. There is a public end point in the positive domain and the standard positive domain. The following elementary dependence functions can be utilized:

$$k_1(x_1) = \begin{cases} \dfrac{\rho(x_1, x_{01}, X_1)}{D(x_1, x_{01}, X_{01}, X_1)}, & D(x_1, x_{01}, X_{01}, X_1) \neq 0, x_1 \in X_1 \\[2mm] -\rho(x_1, x_{01}, X_{01}) + 1, & D(x_1, x_{01}, X_{01}, X_1) = 0, x_1 \in X_{01} \\[2mm] 0, & D(x_1, x_{01}, X_{01}, X_1) = 0, x_1 \notin X_{01}, x_1 \in X_1 \\[2mm] \dfrac{\rho(x_1, x_{01}, X_1)}{D(x_1, x_{01}, X_1, \hat{X}_1)}, & D(x_1, x_{01}, X_1, \hat{X}_1) \neq 0, x_1 \in \mathfrak{R} - X_1 \end{cases} .$$

Because $x_{01} = 150 = a_{01}$, the left extension distance formula can be utilized.

$$\rho(x_1, a_{01}, X_1) = \begin{cases} a_{01} - x_1, & x_1 < a_{01} \\ a_{01} - b_1, & x_1 = a_{01} \\ x_1 - b_1, & x_1 > a_{01} \end{cases}$$

$$= \begin{cases} 150 - x_1, & x_1 < 150 \\ -80, & x_1 = 150 \\ x_1 - 230, & x_1 > 150 \end{cases}$$

$$\rho(x_1, a_{01}, X_{01}) = \begin{cases} a_{01} - x_1, & x_1 < a_{01} \\ a_{01} - b_{01}, & x_1 = a_{01} \\ x_1 - b_{01}, & x_1 > a_{01} \end{cases}$$

$$= \begin{cases} 150 - x_1, & x_1 < 150 \\ -50, & x_1 = 150 \\ x_1 - 200, & x_1 > 150 \end{cases}$$

$$D(x_1, a_{01}, X_{01}, X_1) = \rho(x_1, a_{01}, X_1) - \rho(x_1, a_{01}, X_{01})$$

$$\hat{D}(x_1, a_{01}, X_1, \hat{X}_1) = \rho(x_1, a_{01}, \hat{X}_1) - \rho(x_1, a_{01}, X_1).$$

Thus, the dependence degrees concerning the costs of the three creative ideas respectively are:

$$k_{11}(170) = \frac{\rho(170, 150, X_1)}{D(170, 150, X_{01}, X_1)} = \frac{-60}{-230 + 200} = 2,$$

$$k_{12}(180) = \frac{\rho(180, 150, X_1)}{D(180, 150, X_{01}, X_1)} = \frac{-50}{-230 + 200} \approx 1.67,$$

$$k_{13}(190) = \frac{\rho(190, 150, X_1)}{D(190, 150, X_{01}, X_1)} = \frac{-40}{-230 + 200} \approx 1.33.$$

Upon standardizing the above dependence functions, the standardized dependence degrees can be obtained:

$$K_{11} = \frac{k_{11}(170)}{\left| k_{11}(170) \right|} = 1, K_{12} = \frac{k_{12}(180)}{\left| k_{11}(170) \right|} = 0.835, K_{13} = \frac{k_{13}(190)}{\left| k_{11}(170) \right|} = 0.667.$$

The dependence function of the aesthetic degree: The aesthetic degree is a subjective evaluation characteristic, an as such it varies from one person to another. According to the market survey, this characteristic is acceptable with GOOD or above. Therefore, the following discrete dependence function can be established:

$$K_2(x_2) = \begin{cases} 1, & x_2 = \text{perfect} \\ 0.5, & x_2 = \text{very good} \\ 0, & x_2 = \text{good} \\ -0.5, & x_2 = \text{ordinary} \\ -1, & x_2 = \text{bad} \end{cases}$$

This function is utilized to obtain three standardized dependence degrees: $K_{21} = 0, K_{22} = 0.5, K_{23} = 1$.

The dependence function of technology innovating degree: If the technology innovation is incremental and can be realized, the creative idea is considered as satisfying the given requirements. Moreover, the discrete dependence functions can also be utilized to measure the degree of satisfying requirements:

$$K_3(x_3) = \begin{cases} 1, & x_3 = \text{revolution of technology system} \\ 0.9, & x_3 = \text{revolution of technology - economic paradigm} \\ 0.7, & x_3 = \text{fundamental innovation} \\ 0.5, & x_3 = \text{perfect level of incremental innovation} \\ 0.3, & x_3 = \text{good level of incremental innovation} \\ 0.2, & x_3 = \text{average level of incremental innovation} \\ 0, & x_3 = \text{ordinary level of incremental innovation} \\ -1, & x_3 = \text{no innovation} \end{cases}$$

Upon analysis, the technology innovation of the three ideas belong to the category of incremental innovations at different degrees and are under the category of fundamental innovations. Their dependence degrees are $K_{31} = 0.2, K_{32} = 0.3, K_{33} = 0.5$, respectively.

Calculate the comprehensive superiority: The comprehensive superiority of creative ideas S_1, S_2, and S_3 can be calculated with the weighted sum of the dependence degrees:

$$C(S_j) = (\alpha_1, \alpha_2, \alpha_3) \begin{bmatrix} K_{1j} \\ K_{2j} \\ K_{3j} \end{bmatrix} = \sum_{i=1}^{3} \alpha_i K_{ij}, \ j = 1, 2, 3.$$

Therefore, the superiorities of the three creative ideas can be calculated, respectively, as follows:

$$C(S_1) = \alpha_1 \cdot K_{11} + \alpha_2 \cdot K_{21} + \alpha_3 \cdot K_{31} = 0.38$$

$$C(S_2) = \alpha_1 \cdot K_{12} + \alpha_2 \cdot K_{22} + \alpha_3 \cdot K_{32} = 0.52$$

$$C(S_3) = \alpha_1 \cdot K_{13} + \alpha_2 \cdot K_{23} + \alpha_3 \cdot K_{33} = 0.70.$$

Creative idea S_3 has the greatest superiority, so this is the option that should be selected.

Thinking and Exercises

1. Analyze the notebook you often use with the extensible innovation four-step method and generate superior product ideas.

2. For school students, it is not an easy task to dry their clothes and shoes in the rainy season in southern China because a clothes dryer is not suitable for student dormitories. For this scenario, use the first creation method to obtain multiple new product ideas with the desired function, and then select the superior ideas.

3. Based on the mobile phone you currently use, apply the second creation method to obtain a number of new product ideas, and then select the superior ones.

4. Everyone uses a pen starting in kindergarten. It can be said that the pen is one of the writing implements that accompanies us for the longest time in our lives. What type of pen do you like the most? What do you think of the shortcomings of your favorite pen? Use the third creation method to obtain multiple pen ideas and select superior ones.

5. Analyze one of your professional products or tools. Try to use one of the three creation methods to obtain new product ideas. Then refine one of them as a new product design to apply for a patent.

References

Cai W. Extension set and non-compatible problems. *Journal of Science Exploration*, 1983, (1): 83–97.

Cai W. Extension set and non-compatible problems. In *Advances in Applied Mathematics and Mechanics in China*. Peking: International Academic Publishers, 1990, 1–21.

Cai W. *Matter Element Model and Its Application*. Beijing: Publishing House of Document of Science and Technology, 1994.

Cai W, Yang CY, Lin WC. *Extension Engineering Methods*. Beijing: Science Press, 2003.

Cai W. Extension Theory and Its Application. *Chinese Science Bulletin*, 1999, 44 (17): 1538–1548.

Yang CY, Cai W. *Extension Engineering*. Beijing: Science Press, 2007.

Yang CY, Zhang YJ. *Extension Planning*. Beijing: Science Press, 2002.

Cai W, Yang CY, He B. *Preliminary to Extension Logic*. Beijing: Science Press, 2003.

Wu WJ, etc. The Appraisal of "Extension Theory and Its Applications." http://extenics.gdut.edu.cn/info/1030/1131.htm.

Yang CY. The Methodology of Extenics. In: *Extenics: Its Significance in Science and Prospects in Application*. Beijing: Proceedings of the 271th Symposium of Xiangshan Science Conference 2005, 12: 35–38.

Yang CY, Cai W. *Extenics*. Beijing: Science Press, 2014.

Cai W, Yang CY. Basic Theory and Methodology on Extenics. *Chinese Science Bulletin*, 2013, 58 (13): 1190–1199.

Yang CY, Cai W. *Extenics: Theory, Method and Application*. Beijing: Science Press; Columbus: The Educational Publisher, 2013.

Zhao YW, Su N. *Extension Design*. Beijing: Science Press, 2010.

Yang CY, Li XS. Research progress in extension innovation method and its applications. *Industrial Engineering Journal*, 2012, 15 (1): 131–137.

Yang CY, Cai W. Extenics and intelligent processing of contradictory problems. *Science & Technology Review*, 2014, 32 (36): 15–20.

Yang CY, Cai W. Generating creative ideas for production based on Extenics. *Journal of Guangdong University of Technology*, 2016, 33 (1): 12–16.

Yang CY. Overview of Extension Innovation Methods. In *Communications in Cybernetics, Systems Science and Engineering*. London: CRC Press/Balkema, Taylor & Francis Group, 2013: 11–19.

Liao YQ, Yang CY, Li WH. Extension Innovation Design of Product Family Based on Kano Requirement Model. *Procedia Computer Science*, 2015, 55: 268–277.

Qi NN, Yang CY. Product Conceptual Design Based on Third Creative Method of Extenics. *Mathematics in Practice and Theory*, 2015, 45 (5): 226–238.

Qi NN, Yang CY, Tang L. Multilevel Superiority Evaluation Method and Its Application in Product Purchase Guide. *Journal of Liaoning Technical University (Nature Science)*, 2016, 35 (11): 1351–1358.

Yang CY, Luo LW. Application of Extension Innovation Method in Product Design. *Packaging Engineering*, 2016, 37 (14): 7–10.

Zhao YW, Zhou JQ, Hong HH, et al. Overview and Prospects of Extension Design Methodology. *Computer Integrated Manufacturing Systems*, 2015, 21 (5): 1157–1167.

Kim WC, Mauborgne R. *Blue Ocean Strategy*. Boston: Harvard Business Review, 2015.

Index